ROTARY DRILL

Drill String and Drill Collars

Unit I, Lesson 3
First Edition

Formerly *The Drill Stem*

▼
▼
▼

By S. T. Horton

Published by

PETROLEUM EXTENSION SERVICE
Division of Continuing Education
The University of Texas at Austin
Austin, Texas

in cooperation with

INTERNATIONAL ASSOCIATION
OF DRILLING CONTRACTORS
Houston, Texas

1995

Library of Congress Cataloging-in-Publication Data

Horton, Sheryl T., 1948–
Drilling string and drill collars / by S. T. Horton. — 1st ed.
p. cm. — (Rotary drilling series ; unit I, lesson 3)
"In cooperation with International Association of Drilling Contractors, Houston, Texas."
ISBN 0-88698-168-9
1. Drill pipe. 2. Oil well drilling rigs—Equipment and supplies. I. University of Texas at Austin. Petroleum Extension Service. II. International Association of Drilling Contractors. III. Title. IV. Series.
TN871.5.H67 1994
622'.3381—dc20 93-14987
CIP

First Edition published 1995
Second Impression 1997
Printed in the United States of America

Catalog no. 2.103101
ISBN 0-88698-168-9

Contents

Figures

Foreword

▼
▼
▼

For many years, the Rotary Drilling Series has oriented new personnel and further assisted experienced hands in the rotary drilling industry. As the industry changes, so must the manuals in this series reflect those changes.

The revisions to both text and illustrations are extensive. In addition, the layout has been "modernized" to make the information easy to get; the study questions have been rewritten; and each major section has been summarized to provide a handy comprehension check for the student.

PETEX wishes to thank industry reviewers—and our readers—for invaluable assistance in the revision of the Rotary Drilling Series. Also, we wish to thank the International Association of Drilling Contractors (IADC) for their endorsement of the project. On the PETEX staff, Deborah Caples designed the layout; Doris Dickey proofread innumerable versions; and Ron Baker served as content editor for the entire series. Kathy Bork did her usual superlative job in editing the material.

Although every effort was made to ensure accuracy, this manual is intended to be only a training aid; thus, nothing in it should be construed as approval or disapproval of any specific product or practice.

Sheryl Horton

Acknowledgments

Special thanks to Ken Fischer, director, Committee Operations, International Association of Drilling Contractors, who reviewed this manual and secured other reviewers, and to Kathy Glass, who kept track of the review process. John Altermann, Reading & Bates Drilling Company; Tom Smith, Smith Consulting Services; Jim Arnold, Salem Investment; and Joey Hopewell, Delta Drilling Company, provided invaluable suggestions on the content and language. Without their assistance, this book could not have been written. In addition, special thanks to Jonell Clardy, who clarified the sometimes diffiicult text with wonderful new drawings, and to Terry Gregston, for her excellent black and white photographs.

Sheryl Horton

Units of Measurement

Throughout the world, two systems of measurement dominate: the English system and the metric system. Today, the United States is almost the only country that employs the English system.

The English system uses the pound as the unit of weight, the foot as the unit of length, and the gallon as the unit of capacity. In the English system, for example, 1 foot equals 12 inches, 1 yard equals 36 inches, and 1 mile equals 5,280 feet or 1,760 yards.

The metric system uses the gram as the unit of weight, the metre as the unit of length, and the litre as the unit of capacity. In the metric system, for example, 1 metre equals 10 decimetres, 100 centimetres, or 1,000 millimetres. A kilometre equals 1,000 metres. The metric system, unlike the English system, uses a base of 10; thus, it is easy to convert from one unit to another. To convert from one unit to another in the English system, you must memorize or look up the values.

In the late 1970s, the Eleventh General Conference on Weights and Measures described and adopted the Système International (SI) d'Unités. Conference participants based the SI system on the metric system and designed it as an international standard of measurement.

The *Rotary Drilling Series* gives both English and SI units. And because the SI system employs the British spelling of many of the terms, the book follows those spelling rules as well. The unit of length, for example, is *metre*, not *meter*. (Note, however, that the unit of weight is *gram*, not *gramme*.)

To aid US readers in making and understanding the conversion to the SI system, we include the following table.

Metric Conversion Factors

Quantity or Property	English Units	Multiply English Units By	To Obtain These SI Units
Length, depth, or height	inches (in.)	25.4	millimetres (mm)
	feet (ft)	2.54	centimetres (cm)
	yards (yd)	0.3048	metres (m)
	miles (mi)	0.9144	metres (m)
		1.61	kilometres (km)
Hole and pipe diameters, bit size	inches (in.)	25.4	millimetres (mm)
Drilling rate	feet per hour (ft/h)	0.3048	metres per hour (m/h)
Weight on bit	pounds (lb)	0.445	decanewtons (dn)
Nozzle size	32nds of an inch	0.794	millimetres (mm)
Volume	barrels (bbl)	0.159	cubic metres (m^3)
		159	litres (L)
	gallons per stroke (gal/stroke)	0.00379	cubic metres per stroke (m^3/stroke)
	ounces (oz)	29.57	millilitres (mL)
	cubic inches ($in.^3$)	16.387	cubic centimetres (cm^3)
	cubic feet (ft^3)	28.3169	litres (L)
		0.0283	cubic metres (m^3)
	quarts (qt)	0.9464	litres (L)
	gallons (gal)	3.7854	litres (L)
	gallons (gal)	0.00379	cubic metres (m^3)
Pump output and flow rate	gallons per minute (gpm)	0.00379	cubic metres per minute (m^3/min)
	gallons per hour (gph)	0.00379	cubic metres per hour (m^3/h)
	barrels per stroke (bbl/stroke)	0.159	cubic metres per stroke (m^3/stroke)
	barrels per minute (bbl/min)	0.159	cubic metres per minute (m^3/min)
Pressure	pounds per square inch (psi)	6.895	kilopascals (kPa)
		0.006895	megapascals (MPa)
Temperature	°Fahrenheit (°F)	$\frac{°F - 32}{1.8}$	°Celsius (°C)
Mud weight	pounds per gallon (ppg)	119.82	kilograms per cubic metre (kg/m^3)
	pounds per cubic foot (lb/ft^3)	16.0	kilograms per cubic metre (kg/m^3)
Mass (weight)	ounces (oz)	0.0353	grams (g)
	pounds (lb)	453.59	grams (g)
		0.4536	kilograms (kg)
	tons (tn)	0.9072	tonnes (t)
Pressure gradient	pounds per square inch per foot (psi/ft)	22.621	kilopascals per metre (kPa/m)
Funnel viscosity	seconds per quart (s/qt)	1.057	seconds per litre (s/L)
Power	horsepower (hp)	0.7	kilowatts (kw)
Area	square inches ($in.^2$)	6.45	square centimetres (cm^2)
	square feet (ft^2)	0.0929	square metres (m^2)
	square yards (yd^2)	0.8361	square metres (m^2)
	square miles (mi^2)	2,589.988	square kilometres (km^2)
	acre (ac)	0.40	hectare (ha)
Drilling line wear	ton-miles (tn•mi)	14.317	megajoules (MJ)
Torque	foot-pounds (ft•lb)	1.3558	newton metres (N•M)

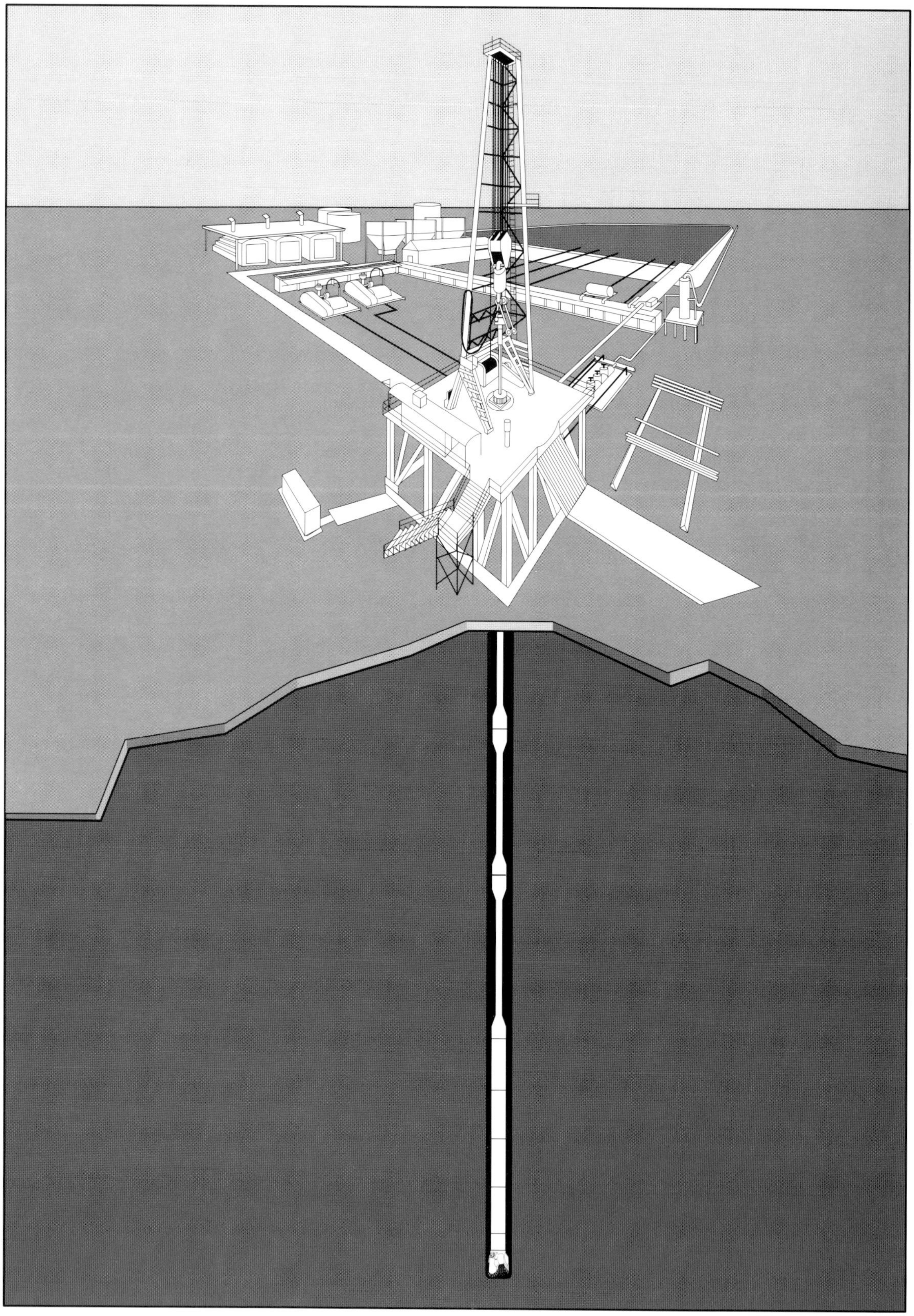

Introduction

▼
▼
▼

In rotary drilling, a crew rotates a bit that drills a hole through the earth in search of oil or gas. The crew attaches the bit to a hollow length of pipe that serves two purposes: it provides the weight to make the bit dig into the earth's formations, and it provides a passageway to circulate a fluid—drilling mud—to the bit as it rotates. This drilling mud cools and lubricates the bit and carries the rock cuttings from the bottom of the hole to the surface (see fig. 1). Surface equipment removes the cuttings and recirculates the clean mud back down the pipe. As the crew drills deeper into the earth, crew members add more pipe to that which is connected to the bit. Before a well is completed, this drill stem may be thousands of feet long.

Because the drill stem serves two purposes, it consists of, among other things, two basic types of pipe: the drill string and the drill collars. These two types of pipe are similar in that they are hollow lengths joined together to make one long conduit from the surface to the bottom of the hole. They are constructed somewhat differently, however, because they fulfill different functions.

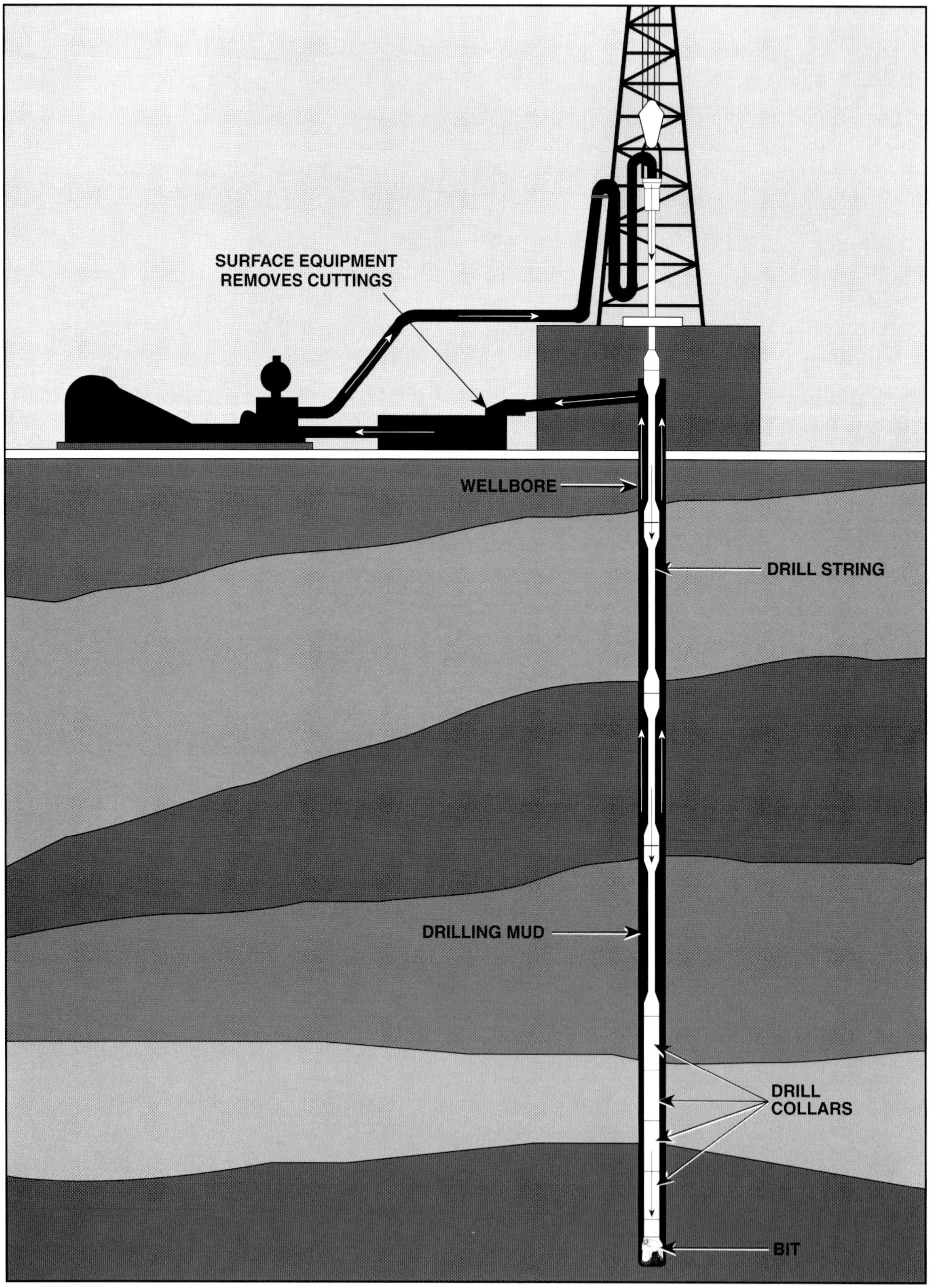

Figure 1. The drill stem and circulating system

Drill Collars

Drill collars are heavy-walled metal tubes, usually made of steel, that the crew positions directly above the bit in order to guide, stabilize, and put weight on it (commonly referred to as WOB—weight on bit) and to make hole. Manufacturers cut threads into each end of a drill collar so that it can be joined with another collar. The end with the threads cut on the inside, the female end, is called the *box*, and the end with the threads cut on the outside, the male end, is called the *pin* (fig. 2). The number of drill collars screwed together and placed on top of the bit depends on the weight needed to make hole efficiently, the type of formations to be drilled through, the weight of each drill collar, and other variables. After the crew attaches the required number of drill collars to the bit, crew members attach additional pipe to continue the conduit from the top of the drill collars to the surface; they use the drill string for this purpose.

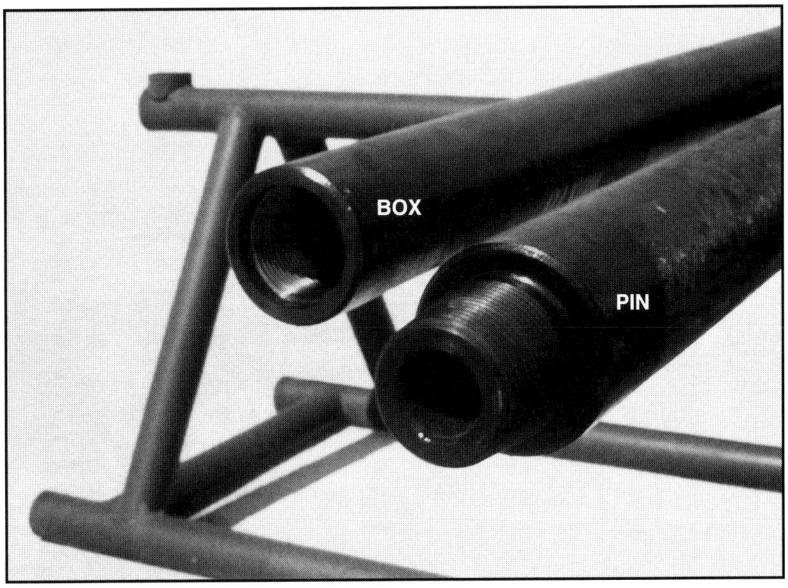

Figure 2. Box and pin ends of collars

Drill String

Drilling crews use drill string to transmit the rotation of the rotary table or top drive to the bit and to serve as a conduit for the drilling fluid. Though the drill string completes the connection between the drill collars and the surface, the crew does not use it to put weight on the bit. Since the drill string is not used to put WOB (unless drilling horizontally, where it is used in compression to push the bit horizontally through the formation), the pipe body is lighter and much thinner than the drill collars. The metal wall of the drill pipe is usually less than half an inch (12.7 millimetres) thick and too thin for threads to be cut directly into it. Manufacturers, therefore, produce short, thick, heavy pieces of pipe called tool joints (see fig. 3) with threads cut on either the outside or the inside. These thicker-walled joints are welded to each end of the drill pipe so that the crew can join the pipes together (see fig. 1).

The term "drill string" refers to the column of drill pipe with attached tool joints. The "drill stem" refers to both the drill string and the drill collars as well as the swivel, kelly, stabilizers, and various specialty items used for rotary drilling. The drill stem is one of the most expensive, integral pieces of rotary drilling equipment. Consequently, handling it properly and maintaining it in good condition are priorities for any crew.

Figure 3. Tool joints

▼
▼
▼

Drill Collars

Standard Design

Manufacturers design drill collars for guiding, stabilizing, and providing weight on the bit. Drill collars are heavy, thick-walled, metal tubes usually made of steel. They range in weight from 16 to 379 pounds per foot (23.81 to 564.01 kilograms per metre). Most drill collars are round and are 30 or 31 feet (9.14 or 9.45 metres) long. Drilling crews join the required number of collars by stabbing the pin end of one collar into the box end of another and screwing them together (fig. 4).

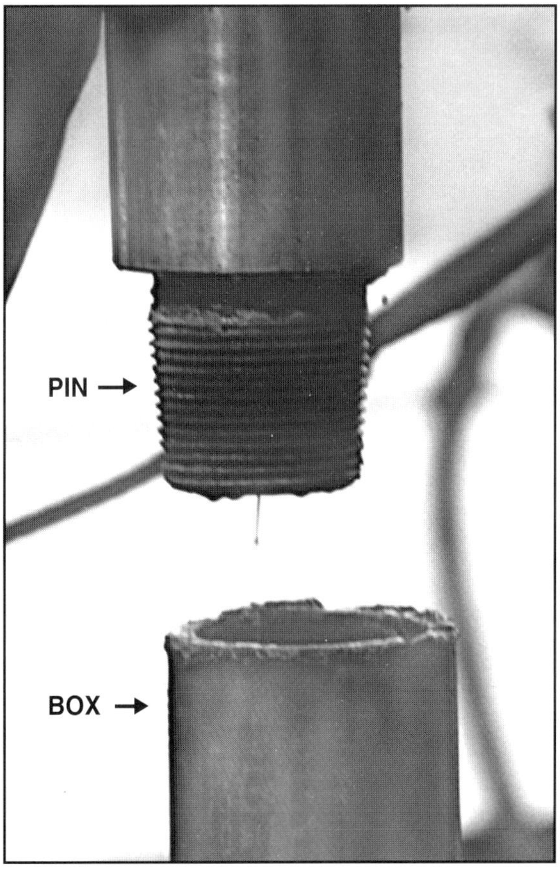

Figure 4. Stabbing the pin of one collar into the box of another

The well planner determines how many drill collars are needed above the bit by taking into consideration how much weight is needed to drill efficiently and how much each drill collar weighs. Generally, the bigger a drill collar's outside diameter, the more it weighs. For example, a 30-foot (9.14-metre) drill collar with an outside diameter (OD) of 6 inches (152.4 millimetres) and an inside diameter (ID), or bore diameter, as it's also called, of 2¼ inches (57.15 millimetres) weighs 2,490 pounds (1,129 kilograms). On the other hand, a 30-foot (9.14-metre) drill collar joint with an OD of 11 inches (279.4 millimetres) and an ID of 3 inches (76.2 centimetres) weighs 8,843 pounds (4,011 kilograms).

Drill collar ID varies from 2 inches (50.8 centimetres) to 3 inches (76.2 centimetres), but the OD ranges from 4 inches (101.6 centimetres) to 11 inches (279.4 centimetres). The well planner selects the correct OD based on the size of the wellbore, and then determines the number of drill collars needed to provide the proper weight-on-bit to make hole.

An important factor to consider when connecting collars is the relationship between the pin and the box. Engineers have learned to compare the bending strength of the pin near its base (near the last engaged pin thread, or LET) to the bending strength of the box near its bottom (near the LET—fig. 5). In good design, the bending strength of the box should be 2.5 times greater than the bending strength of the pin. (For more information on why this ratio is important, see pages 18–22.)

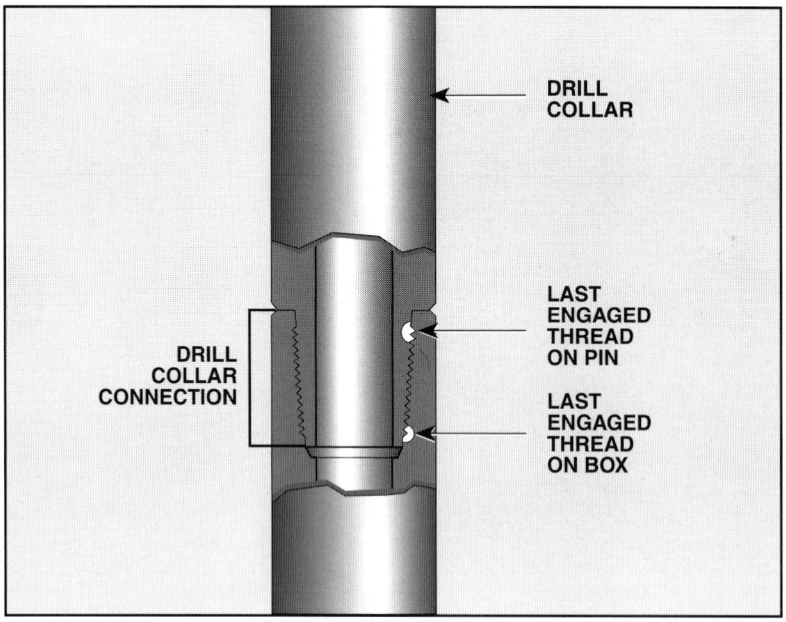

Figure 5. In a drill collar connection, the last engaged thread on the the pin and the box

Special Designs

Spiral Collars

Although most collars are round, engineers design some collars to counteract specific problems that can develop while drilling. For instance, some collars may have spiral grooves machined into the outside surface (fig. 6). Drillers sometimes use spiral drill collars in holes in which the clearance between the drill collar and the wall of the hole is small and in directional drilling, where the collars will likely contact the side of the hole, to minimize *differential sticking*. Collars with spiral grooves help prevent differential sticking by reducing the amount of drill collar surface area that contacts the sides of the hole. Differential sticking is a condition in which the drill stem becomes stuck against the wall of the wellbore because of the difference in pressure between the drilling fluid in the wellbore and that of a permeable formation. The pressure is higher in the wellbore. Thus, the differential pressure forces some of the liquid part of the drilling fluid into the permeable formation. This leaves a layer of concentrated mud solids on the side of the hole. This layer of mud solids is called filter cake. The lower formation pressure creates suction that holds the collar against the side of the hole (fig. 7). If the pressure differential between the wellbore and the formation is sufficient and the clearance between the drill collar and the hole is small, it is easy for the collar to contact the filter cake and get stuck.

Figure 6. Spiral drill collars

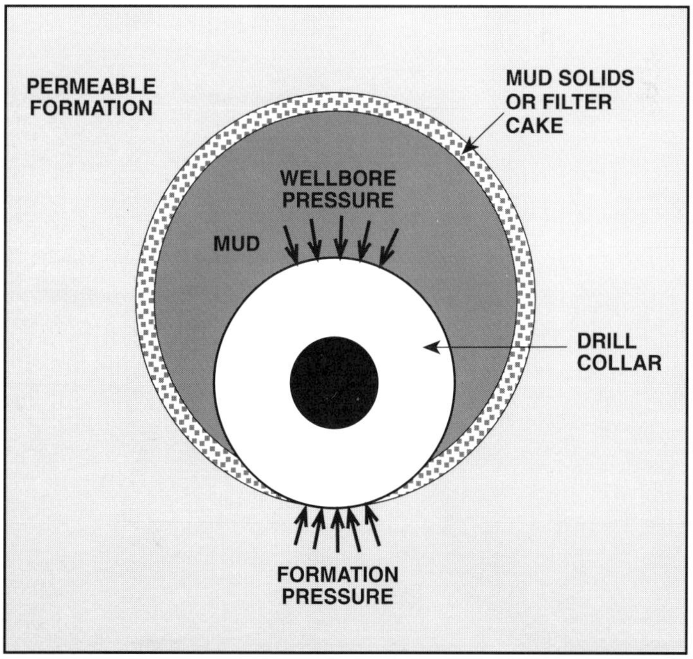

Figure 7. Differential sticking

Square Collars

Drillers use square drill collars (fig. 8) in an effort to keep the wellbore on as vertical a path as possible. In straight-hole drilling, the driller attempts to keep the wellbore as vertical as possible until the target is reached. One phenomenon that assists the driller is the *pendulum effect*, or the tendency of the drill stem to hang in a vertical position because of the force of gravity pulling on the weight of the drill collars. The heavier the drill collars, the greater the pendulum effect. Square drill collars have more steel and are thus stiffer and heavier than round ones. They tend to give, or bend, less than round collars and produce a greater amount of pendulum force. Also, each corner of a square drill collar touches or nearly touches the wall of the hole as the drill stem rotates. This contact (or near contact) acts to stabilize the collar, or keep it stiff and straight. With no wobble, the collar keeps the bit drilling on course.

Figure 8. Square drill collar

Other Special Features

In addition to square and spiral drill collars, manufacturers produce other varieties of drill collar with special features. These special features include such items as recesses for slips and elevators (fig. 9) and hard banding (fig.10).

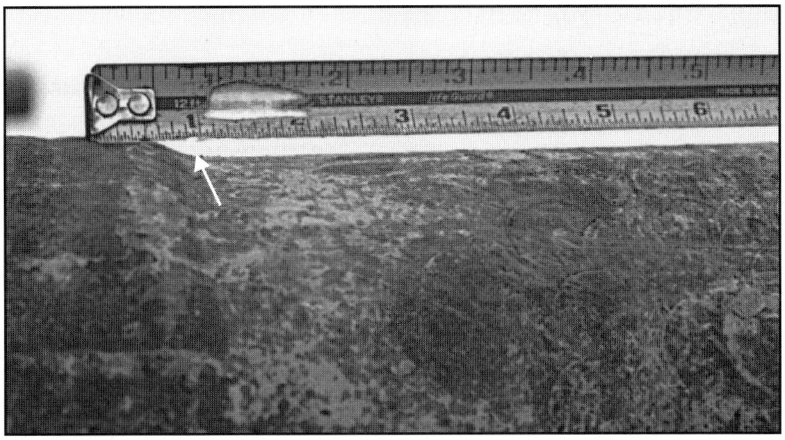

Figure 9. Collar with recess for use with slips and elevators

Figure 10. Hardbanding on collars

Functions

A string of drill collars—including two or three collars in some areas or ten times that many in others—performs the following tasks:

1. It provides weight on the bit for drilling.
2. It maintains weight to keep the drill string from being subjected to buckling forces.
3. It helps provide the pendulum effect which causes the bit to drill a more nearly vertical hole.
4. It helps support and stabilize the bit so that it will drill new hole that is aligned with the hole already drilled.

Weight on Bit

As already noted, drill collars provide weight on the bit in order to make the bit gouge into the earth and make hole. The amount of weight required depends on several factors. The well planner considers the kind of formations being drilled, the diameter of the hole, the kind of bit being used to drill, the buoyant effect of the drilling fluid, the tendency of the hole to deviate from a true vertical path, and other variables. The correct amount of weight on the bit helps make for efficient drilling of the hole. The weight of each drill collar depends on the OD and the ID, or bore, of the pipe. For example, in figure 11, two drill collars have the same ID but different OD. In this case, the drill collar with the larger OD is the heavier collar (per foot or metre).

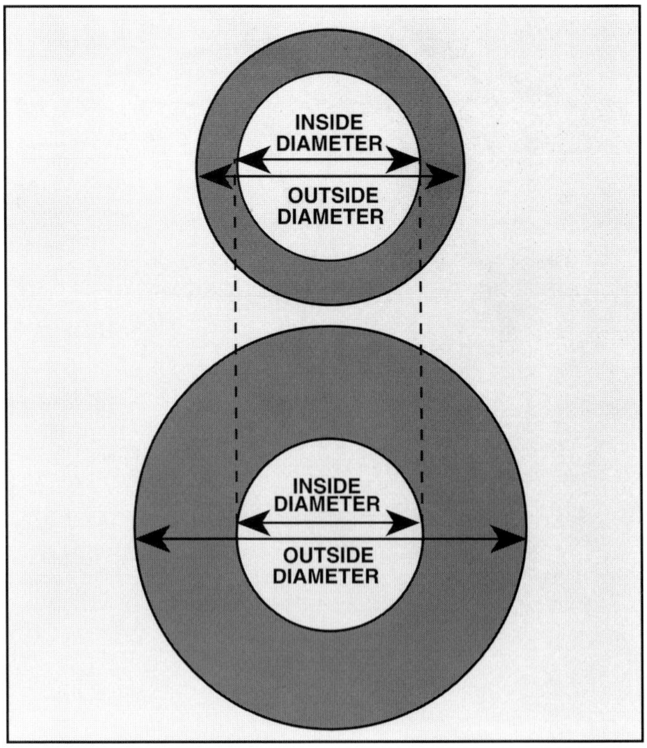

Figure 11. Two collars with the same inside diameters but different outside diameters

The weight or force needed on the bit may be as little as a few thousand pounds (decanewtons) in the case of a small-diameter hole being drilled with a small bit; the amount may rise to 100,000 pounds (44,500 decanewtons) or more in a large-diameter hole being drilled with a large bit.

The buoyant effect of the drilling fluid in the hole must be taken into consideration when determining WOB. Just as a boat floats on water, the drill stem floats in drilling fluid, thereby reducing the amount of weight on the bit. The denser the drilling fluid, the greater the buoyant effect, so more drill collars are needed to provide the same weight on the bit. Drilling crews often use additional collars to offset this buoyancy. For example, a common range of WOB is from 30,000 pounds (13,350 decanewtons) to 60,000 pounds (26,700 decanewtons). A common drilling fluid used weighs 10 pounds per gallon (ppg) (1,198.20 kilograms per cubic metre). A 10-ppg (1,198.20-kilogram-per-cubic metre) drilling fluid has a buoyancy factor of 0.847. To account for the buoyant effect of the drilling fluid, the operator takes the required WOB and divides it by the buoyancy factor of the mud to arrive at the air weight of the drill collars. For example, suppose that the operator determines that the WOB should be 60,000 pounds (26,700 decanewtons). Further suppose that the crew is drilling with a 10-ppg (1,198.20-kilograms-per-cubic-metre) drilling fluid. To determine the weight needed to overcome buoyancy, the operator would divide 60,000 by 0.847. The quotient (70,838) represents the air weight of the drill collars needed to put 60,000 pounds of WOB in 10-ppg drilling fluid.

In SI units the operator would divide 26,700 decanewtons by 0.847 to get 31,523.0. Thus, to overcome buoyancy in 1,198.20-kilograms-per-cubic-metre drilling fluid and put 26,700 decanewtons of force on the bit, the collar section in air would have to exert 31,523.0 decanewtons.

The amount of deviation in the wellbore must also be taken into consideration when determining WOB. The more the hole deviates from vertical, the greater is the reduction in effective WOB. For example, in a hole that deviates from vertical by 60°, the reduction in effective WOB can be as much as 50%. So, depending on how much the hole deviates from vertical, the well planner must increase the number of drill collars to provide the required WOB.

Holding the Drill String Straight

In addition to putting weight on the bit to make hole efficiently, drilling crews must also use the weight of the collars to keep the drill string above the collars from bending or buckling. If the drill string is kept in tension, it will not bend or buckle. The drill string is kept in tension by two opposing forces—the weight of the drill collars pulling on it from below and the drawworks and drilling line pulling on it from the surface (fig. 12).

Under normal conditions, the drill pipe string wears more rapidly near the center of each joint. Bending or buckling increases the metal fatigue in the center of the joint—that is, bending increases the stress factors in the pipe's weakest area. Bending also forces the drill string to rub against the wall of the hole, which abrades the sides of the pipe.

Sometimes the driller uses extra weight on the collars to make sure that the drill string stays straight. When using additional weight to hold the drill string straight, the driller usually uses drill collars having a buoyed weight 10% to 30% greater than the amount needed just to make hole. Providing this extra weight will help keep the drill string relatively straight as it is rotated.

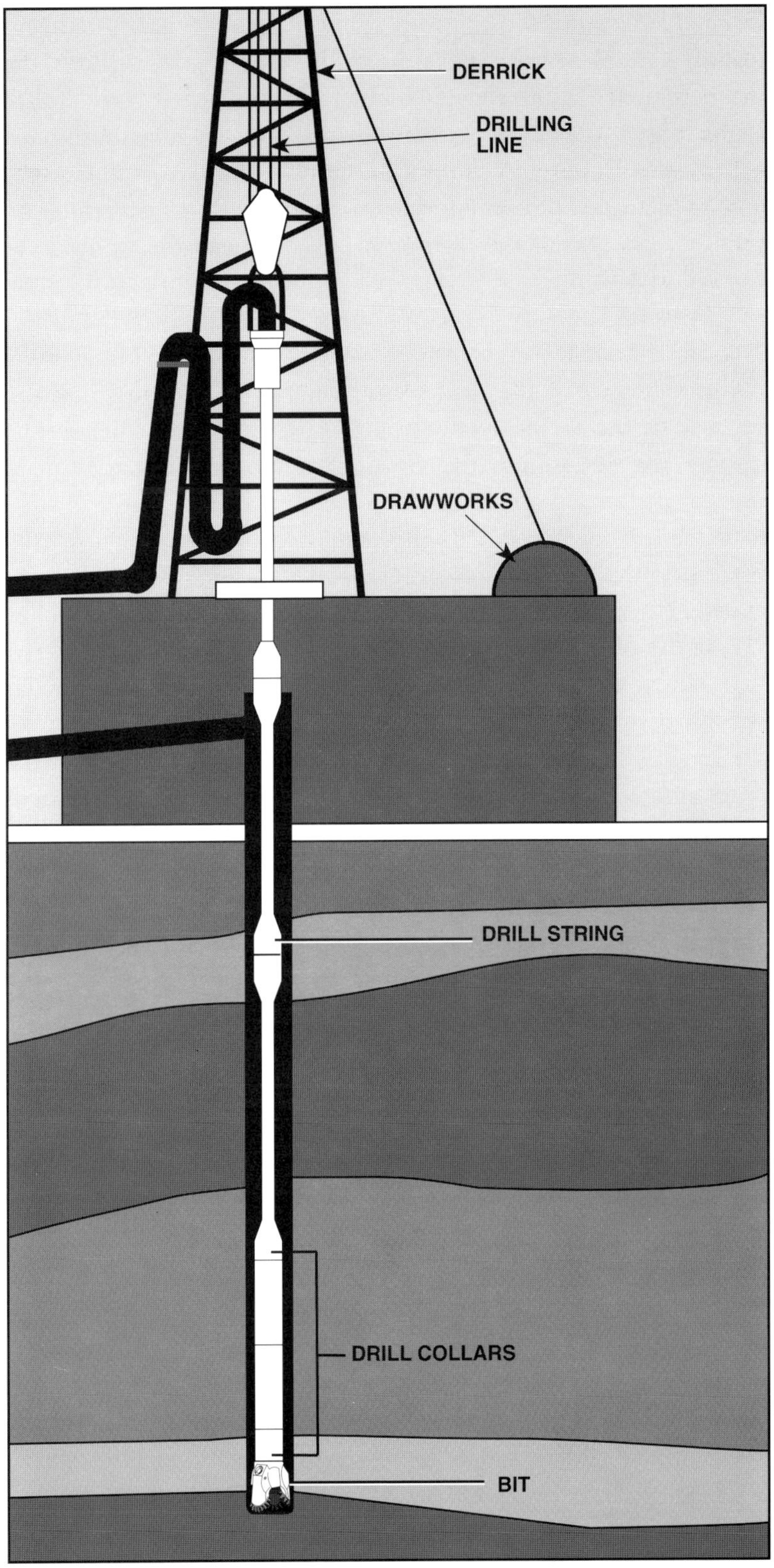

Figure 12. The drill string is kept in tension by two opposing forces—the weight of the collars and the pull of the drawworks and the drilling line.

Providing a Pendulum Effect

Drill collar weight also provides a pendulum effect to cause the bit to drill a more nearly vertical hole. In straight-hole drilling, the most efficient drilling is that which produces a nearly vertical path to the target. Any deviation from vertical tends to increase the cost of the well. Of course, no hole is absolutely vertical from the surface to completion. Different forces cause the hole to deviate from vertical. The job of the drilling crew is to keep the wellbore as vertical as possible.

As stated previously, one phenomenon that assists the driller in straight-hole drilling is the *pendulum effect*. The pendulum effect is the tendency of the drill stem to hang in a vertical position because of the force of gravity pulling on the drill collars. The heavier the pendulum, the stronger is its tendency to remain vertical and the greater is the force needed to cause the drill stem to deviate from vertical. If the drill stem deviates from a true vertical path, the force of gravity tends to pull it back toward vertical (fig. 13). The bit and drill collars tend to drill on the low side of the hole as they seek a more nearly vertical path.

The length of the pendulum is that section of the drill collar string between the bit and the lowest-point tangent to the side of the hole. The lowest-point tangent to the side of the hole is the point at which the drill collars first touch the side of the hole or at which the first stabilizer touches the side of the hole. The drilling crew tries to keep this point as high as practical so that the pendulum will be as long as possible. The longer the pendulum, the greater is the tendency for the drill stem to seek a vertical position. There is a limit, however, on the spacing between the bit and the first stabilizer. If the distance is too great, the drill collars may buckle. Tables are available to assist in selecting the appropriate spacing in relation to hole size, drill collar OD and ID, and the angle of the hole.

The weight, the length, and the stiffness of the pendulum are important in straight-hole drilling. A stiff drill collar assembly has a higher point of tangency than a limber one. To achieve both weight and stiffness, crews use large, heavy drill collars. Undersized and relatively limber drill collars, with a large clearance between the OD of the collar and the wall of the hole, make it harder to keep the hole vertical. Relatively limber drill collar strings tend to bend or flex near the bit; this bending shortens the pendulum considerably and increases the deviation from vertical.

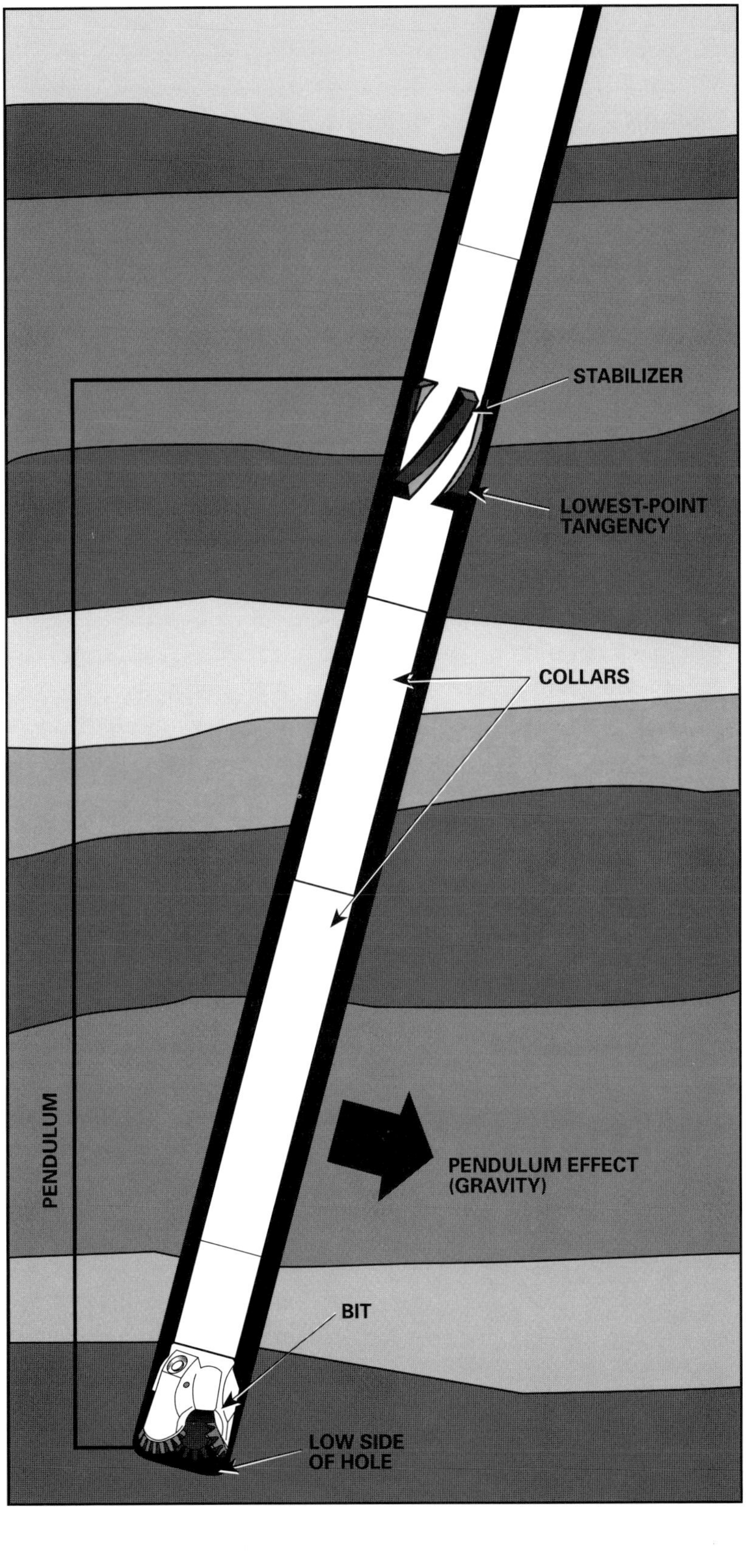

Figure 13. Gravity pulls the bit to the low side of the hole, producing a pendulum effect.

Stabilizing the Bit

Drilling crews use large, heavy, round drill collars with stabilizers to help stabilize the bit when drilling. They use these collars and stabilizers to create what is called a *packed bottomhole assembly* (BHA) (fig. 14). This term refers to the fact that the round drill collars with stabilizers in the lower part of the drill collar assembly are almost the same OD as the bit. In addition, the stabilizers are no more than ⅛ inch (3.18 millimetres) smaller in diameter than the hole. In a packed BHA composed of a bit and round drill collars with three or more stabilizers above it, the crew places the first stabilizer immediately above the bit, followed by a short, large-diameter drill collar and a second stabilizer. The third stabilizer is located farther up in the drill collar string.

The use of a packed bottomhole assembly is sometimes referred to as a *gun-barrel approach* because when the crew uses this assembly, the bit tends to drill straight ahead in spite of forces trying to prevent it from doing so. In other words, the assembly assures that the new wellbore is being drilled in line with the hole that has already been drilled; it is as straight as the barrel of a gun. Using a packed bottomhole assembly, or the gun-barrel approach, keeps the bit from wobbling, or walking, on the bottom of the hole. This approach also keeps the pressure on the cutting structure of the bit uniform, thereby forcing the bit to cut evenly through the formation. A bit that is not connected to a packed bottomhole assembly may drill an oversized hole, produce unusual bit wear, and slow the rate of penetration. In general, bits drill faster and last longer when well stabilized.

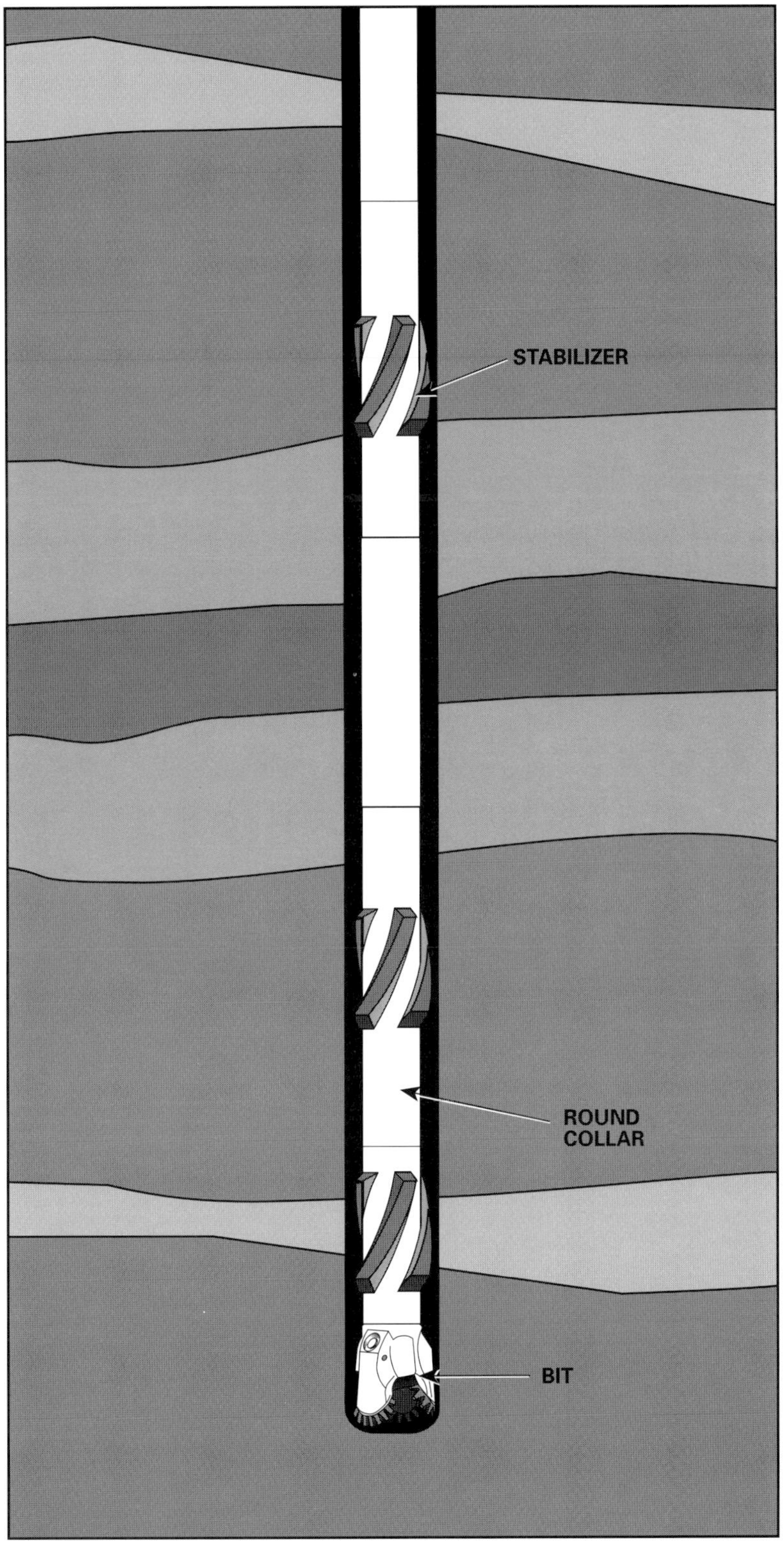

Figure 14. A packed bottomhole assembly

Failures

Although drill collars are heavy and stiff, they can fail. A single 8-inch (203.2 millimetre) drill collar on the rack may appear to be so heavy and stiff that it cannot bend, but when three of the collars are screwed together and standing in the derrick, a definite bend is noticeable. In the hole, twenty of these collars may be stacked into a long and relatively limber column that bends when compressed by the weight of the entire column. During drilling, rotational force puts even more stress on the collars. In fact, the tension and compression put on collars during drilling can result in cracks and breaks.

In any string of drill collars, the metal fibers on the outer side of a bend are under tension while the fibers on the inner side of a bend are under compression (fig. 15). If the collars are rotated one-half turn, the fibers that were under tension are put under compression. In every drill collar rotated in the hole, this continuous movement from tension to compression takes place with each rotation of the drill stem.

In a drill collar, there are two main areas of stress concentration—one in the pin, and one in the box (fig. 5). Drill collars usually fail in the threaded connections because most of the bending stress is concentrated at this point. If the surface of any drill stem component is cracked, the crack tends to enlarge every time it comes to the tension side of the bend. Such cracks are called fatigue cracks.

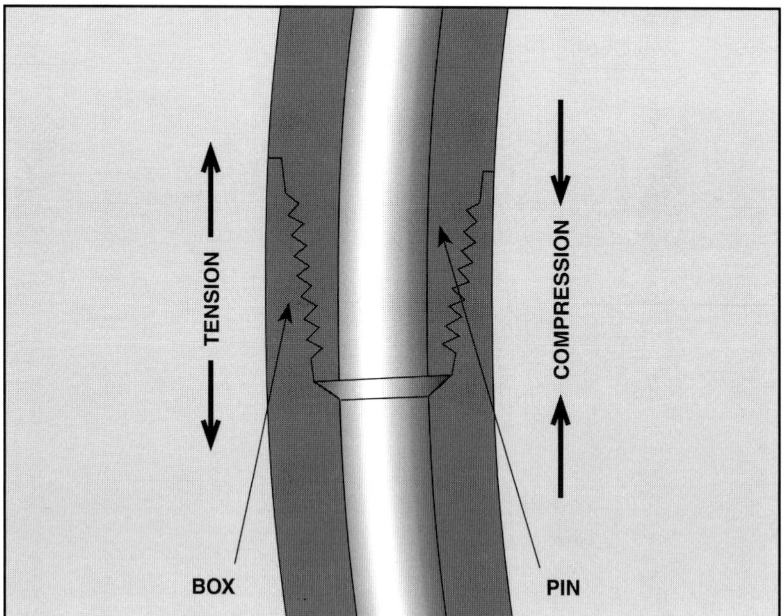

Figure 15. Tension and compression on drill collars during drilling

Making a Good Connection

In order to reduce the number of drill collar failures, crew members should make up a drill collar joint correctly—that is, they should make a *good connection*. They should use the proper torque (the proper amount of turning force needed to screw the two collars together as published in API RP 7G *Recommended Practices for Drill Stem Design and Operating limits*, 15th ed. and Chapter B of the International Association Drilling Contractors (IADC) *Drilling Manual*, 11th ed.), proper procedures, and the correct amount of lubricant, also called thread compound, or dope. In a good connection, the shoulders of the collar support the pin so that it is rigidly held by the box (fig. 16). In a good connection, the only seal between the two joints is made between the shoulders of the pin and the box. The pin and the box serve as a simple screw jack pulling the two shoulders together very tightly. Unlike the threads in tubing and casing, which produce a continuous seal, the threads of the collars do not form a pressure seal. In a good connection, the shoulders on the collars must be connected with adequate *preload*, or proper makeup torque, to produce a seal that will hold even under the tension and compression put on the drill collars during drilling.

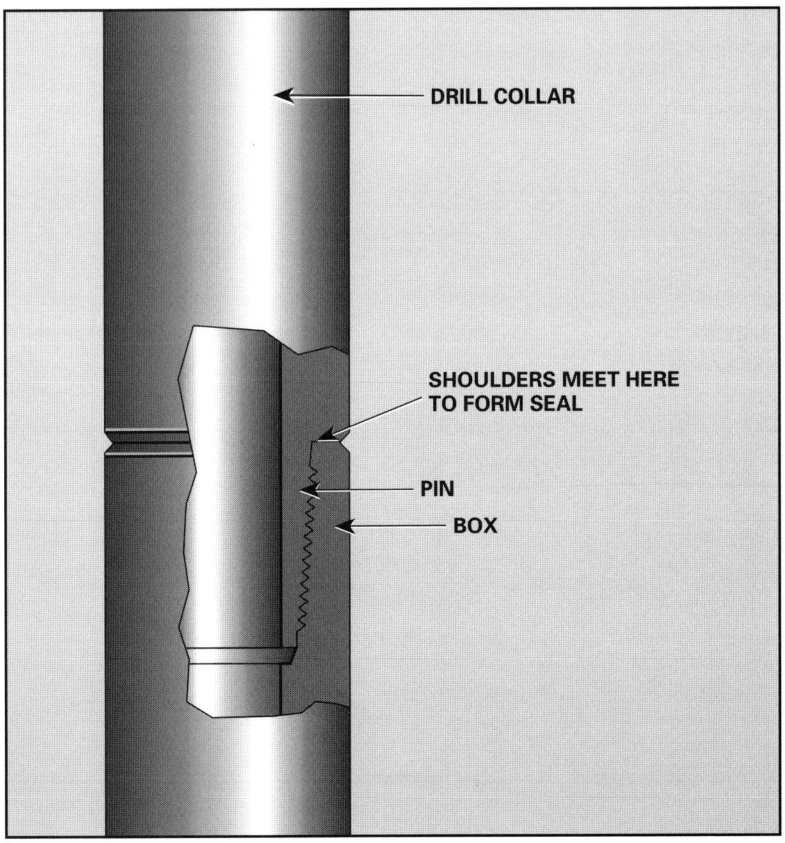

Figure 16. In a drill collar connection, the shoulders form the seal between the two collars.

Failures with Good Connections

Even when the crew makes up the drill collars correctly, the drill collars can sometimes fail. The weakest section in a good connection (fig. 5) may be the narrow cross section in the bottom of a badly worn box, or a box with low bending strength near the end of the pin. As mentioned earlier, the bending strength of the box should be 2½ times greater than the bending strength of the pin to withstand the stresses during drilling. During drilling, rotation of the drill collars puts high cyclic stress on the bottom of the box. A small crack may form, which can result in a box failure.

Making a Bad Connection

A crew can make a bad connection by using too little torque, too much torque, an inadequate amount of the proper lubricant, or improper procedures. When crew members use too little torque, tension in drilling can cause an opening between the shoulders of the pin and the box. As the drill collar above the shoulders bends to one side, it puts a severe tension load on one side of the collars and compression on the other. As the drill collar rotates one-half turn, the side that was in compression is now in tension; the side that was in tension is now in compression. If the collars were not made up to the proper torque, this rocking back and forth may cause a leak between the shoulders. Every time the seal breaks, drilling fluid passes across the shoulder (fig. 17).

If drilling fluid passes across the shoulder, three things happen: (1) the lubricant, or dope, between the shoulders washes out; (2) the surface of the shoulders begins to erode; and (3) if enough fluid moves through the seal, the lubricant is washed away from the threads as well as from the shoulders. The crew can recognize a loose joint when breaking out drill collars because the pin is dry (no lubricant is present).

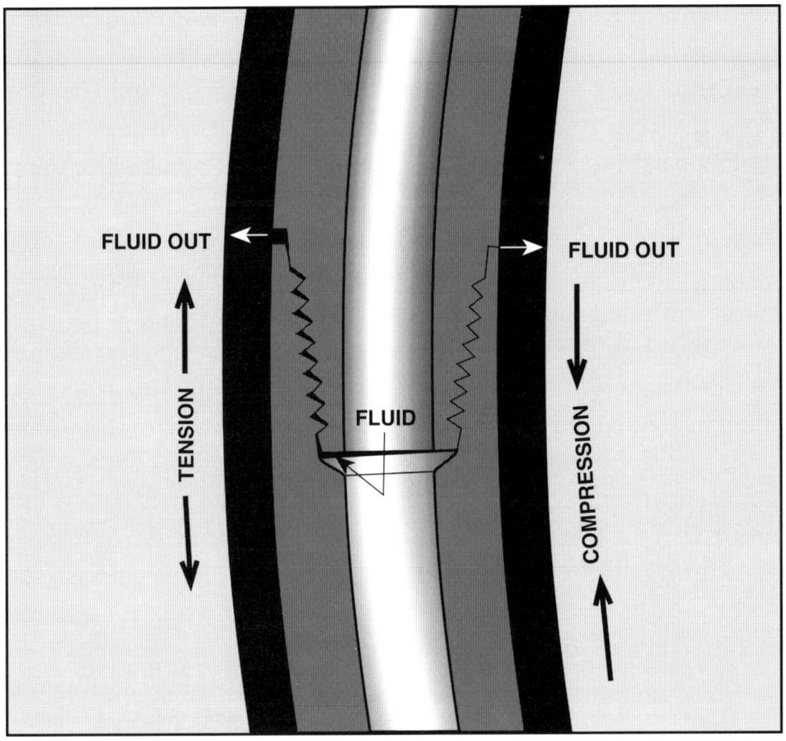

Figure 17. If the seal between collar shoulders is broken, drilling fluid leaks out.

Failures with Bad Connections

When drilling with an undertorqued connection, failures normally occur in the first or second thread near the base of the pin (fig. 5); this thread is often referred to as the last engaged thread, or LET. Occasionally, however, the box fails—not at the bottom, but 1½ to 2 inches (38.1 millimetres to 50.8 millimetres) from the shoulder at about the third or fourth thread. Such failures occur when the box cross sections are relatively weak compared to the pin.

Too little torque causes more failures than too much torque, but too much torque, especially with smaller-sized collars, may result in failures also. Failures caused by overtorquing occur in the weaker member—in the LET near the base of the pin or in the box if the shoulder is small. If the crew makes up the pin and the box together with too much torque, something will break. The tongs break, the pulling line breaks, or the box swells (fig. 18) as the overtorqued pin pushes against it. The pin may actually screw through the box. Sufficient torque must be used to make a good connection, but too much torque causes failure.

The shoulder area can be too small to form a good connection because the pin or box has been improperly machined or because the outside diameter of the box has been severely worn. In either case, the box shoulder cannot support the bending stresses imposed on it by the pin because it has a small contact area. This means that the connection cannot long withstand the forces placed on it during drilling. The high bending stresses on the box near the mouth cause the shoulder to shear off and the box to bell and swallow the pin. (fig. 18).

Figure 18. Swelled box and swallowed pin

When rotating downhole, drill collars tend to overtighten when drilling torques are high—that is, they tend to screw together even more tightly than when they were made up. The result may be a belled box. Slick thread compounds aggravate this problem by allowing the connection to be overtorqued more easily. Box failures of this type are rare if the joint has a good balance between pin size and the outside diameter of the box. If drill collars with narrow shoulders must be used, the driller should make sure the joints are made up to the proper torque, avoid high torques while drilling, and use a high-quality drill collar thread lubricant.

High stress loads on mating threads and shoulders and inadequate or improperly applied thread compounds may lead to *galling*. Galling occurs when two mating steel surfaces, not protected by a film of lubricant (such as metallic zinc or copper), freeze together. When the connection is broken out, the metal is torn apart and the threads and shoulders are damaged (fig. 19). Galling at its worst can destroy one or both makeup shoulders and can wipe out all the threads in a pin or a box. Galling is more frequent on newly machined surfaces.

Figure 19. Galled threads

API-Recommended Torque

API RP 7G lists recommended makeup torques for various sizes of drill collars and types of connections. A common error in determining the makeup torque for a given size and type of drill collar is to fail to take into account the true outside diameter of the drill collar after it has worn somewhat. A new drill collar with API NC 56 connections that is 7½ inches OD × 2 13⁄16 inches (bore) (190.5 millimetres OD × 71.46 millimetres bore) requires a minimum of 48,221 foot-pounds (64,414 newton metres, or N•m) of makeup torque (fig. 20). If it is worn down, however, to a 7¼-inch (184.15 millimetres) OD, the makeup torque should be reduced to a minimum of 40,498 foot-pounds (54,907 N•m).

Another common error is to use tool joint specifications instead of drill collar specifications. An API NC 46 tool joint that is 6 inches OD × 3¼ inches bore (152.4 millimetres OD × 82.55 millimetres bore) requires a minimum of 20,175 foot-pounds (27,353 N•m). The same-sized thread used on a 6 inch × 2 13⁄16-inch bore (152.4 millimetres × 71.44 millimetres) drill collar requires a minimum of 22,426 foot-pounds (30,405 N•m) of makeup torque.

API RECOMMENDED PRACTICE 7G

Table 14—Recommended Make-up Torque for Rotary Shouldered Drill Collar Connections

(1)	(2)	(3)	(4)	(5)	(6)	(7)	(8)	(9)	(10)	(11)	(12)	(13)
Connection			Minimum Make-up Torque, ft-lb									
			Bore of Drill Collar, inches									
Size, in.	Type	OD, in.	1	1¼	1½	1¾	2	2¼	2½	2 13/16	3	3¾
4½	Extra Hole	5¾						*17,738	*17,738	*17,738	*17,738	
API	NC 46	6						*23,422	*23,422	22,426	20,311	
4	API IF	6¼						28,021	25,676	22,426	20,311	
4½	Semi IF	6½						28,021	25,676	22,426	20,311	
5	Dbl. Streamline	6¾						28,021	25,676	22,426	20,311	
4½	Mod. Open											

Source: API RP 7G, 15th ed., January 1, 1995. Reprinted courtesy of the American Petroleum Institute.

RECOMMENDED PRACTICE FOR DRILL STEM DESIGN AND OPERATING LIMITS

Table 10—Recommended Minimum OD and Make-up Torque of Weld-on Type Tool Joints Based on Torsional Strength of Box and Drill Pipe

(1)	(2)	(3)	(4)	(5)	(6)	(7)	(8)	(9)	(10)	(11)	(12)	(13)
Drill Pipe			New Tool Joint Data				Premium Class			Class 2		
Nom Size in.	Nom Wt lb/ft	Type Upset and Grade	Conn.	New OD in.	New ID in.	Make-up Torque ft-lb	Min OD Tool Joint in.	Min Box Shoulder with Eccentric Wear in.	Make-up Torque for Min OD Tool Joint ft-lb	Min OD Tool Joint in.	Min Box Shoulder with Eccentric Wear in.	Make-up Torque for Min OD Tool Joint ft-lb
4	14.00	IU 105	NC 40	5½	2 7/16	18,068 P	5	9/32	12,569	4 29/32	15/64	10,768
	14.00	EU 105	NC 46	6	3¼	20,175 P	5 7/16	7/32	12,813	5 11/32	11/64	10,647
	14.00	IU 105	4 H90	5½	2 13/16	21,224 P	5 3/32	7/32	12,481	5 1/32	3/16	11,065

Source: API RP 7G, 15th ed., January 1, 1995. Reprinted courtesy of the American Petroleum Institute.

RECOMMENDED PRACTICE FOR DRILL STEM DESIGN AND OPERATING LIMITS

Table 14—Recommended Make-up Torque for Rotary Shouldered Drill Collar Connections

(1)	(2)	(3)	(4)	(5)	(6)	(7)	(8)	(9)	(10)
Connection			Minimum Make-up Torque ft-lb						
			Bore of Drill Collar, inches						
Size in.	Type	OD, in.	2¼	2½	2 13/16	3	3¼	3½	3¾
API	NC 56	7¼		*40,498	*40,498	*40,498	*40,498		
		7½		*49,060	*48,221	*45,680	*42,058		
		7¾		*52,115	*48,221	*45,680	*42,058		
		8		*52,115	*48,221	*45,680	*42,058		

Source: API RP 7G, 15th ed., January 1, 1995. Reprinted courtesy of the American Petroleum Institute.

Figure 20. Table: Recommended make-up torque

Care and Handling

Because collars are such an expensive part of the drilling equipment, the crew must keep them in good condition and handle them properly to lengthen their life. Crew members must treat the collars with care not only when they are making them up and breaking them out, but also when they are storing them and moving them about the rig.

Handling the Collars

Drill collars are heavy and awkward to move around the rig. The crew must be very careful to prevent damage to the threaded ends and shoulders. Putting the heavy collars down on unprotected threads or knocking the threads with another collar can easily damage them. Crew members should keep thread protectors (fig. 21) on both ends of the collar when they move the collars around the rack and across the catwalk. Some thread protectors have bails, or handles (fig. 22), which make it easier for crew members to pick up the collar. Regardless of how collars are moved, the thread protectors should be in place first.

When crew members are ready to move the collars through the V-door, they use a lifting sub or nipple (fig. 23) screwed onto the collars if the collars do not have recesses. Unless manufactured with recesses, collars require either subs or nipples to provide a place for the elevators to attach. Before using subs or nipples, however, crew members should check the threads and the shoulder on the subs or nipples to be sure they are clean and that they fit the collar.

Before a drill collar can be made up to another one, crew members should make sure that the threads are clean and dry and that they are not rusted. Keeping the threads clean increases the life of the connections by eliminating any abrasive materials that can damage the threads. Crew members should also inspect the collar to make sure that it is not damaged. A damaged drill collar should never be run into the hole. The drilling crew should regularly inspect the collars to make sure they are in good condition. Even slight damage can cause problems downhole. (For more details on inspecting the drill collars, see *Inspection* in this book.)

Figure 21. Thread protector on collar

Figure 22. Thread protectors with bails (handles)

Figure 23. Lifting sub

Handling the Collar during Makeup and Breakout

After crew members have made sure that a collar is clean and in good condition, they can make it up to the bit or to another collar. They should apply a good coating of thread compound, or dope, to the threads and shoulders. This lubricant, normally a soft metallic component like copper or zinc in a grease base, lubricates and separates mating threads and shoulders in spite of the high stresses put on them during makeup and while drilling. It also makes it easier to screw the two joints of collar together. Crew members should never dilute or thin the thread component because this will reduce the percentage of metallic component and thus increase the chances of galling the threads. Because many cleaning fluids used on the collars can dilute the lubricant, crew members should make sure that the collar threads are completely dry before applying the compound.

The crew should take great care not to damage the threads or shoulder when stabbing the pin into the box. Dropping the pin onto the box shoulder or jarring the threads during makeup can cause severe damage.

After stabbing the pin of one collar into the box of another, crew members hand-tighten the two collars together by "walking in" the one on top while the rotary keeps the lower one stationary. Hand-tightening the collars before making them up to final torque helps preserve the threads. The crew should hand-tighten the collars using the chain tongs rather than the spinning chain because the spinning chain would spin too fast and possibly gall the threads. Also, the collars are so big and heavy that it is easier for the crew to hand-tighten them by holding the top collar with tongs and walking around the rotary turning the top collar until it is screwed into the bottom collar. On some rigs the crew uses a spinning wrench to make up the collars hand-tight. The crew can use the spinning wrench only on collars that are not over 9⅛ inches (242.3 millimetres) in diameter. If the outside diameter of the collar is larger than 9⅛ inches (242.3 millimetres), the wrench will not fit around it.

To make up the drill collars to the final recommended torque, the crew uses the breakout tongs and the cathead. The drill collars are so heavy that they require a great deal of pulling force to be made up to proper torque. The breakout cathead uses a very strong cable attached to the end of the breakout tongs. Cable is stronger than chain and is therefore less likely to break. In addition, the breakout cathead and cable are located well away from the driller. Thus, if there is a break in the cable, the driller will not be in danger. If crew members used the makeup cathead and tongs, they would be using a chain instead of a cable and would be much closer to the driller—too close for safety. Drill collar connections should never be made up or broken out by turning the rotary because the rotary spins the collars with too much force and the collars can be damaged. After crew members have broken a connection using the tongs, they may then use the rotary to finish spinning it out.

After crew members make a connection, they are ready to lower the collar into the hole and attach another one. They attach the elevators to a lifting sub that is made up to the top of the collar that is above the rotary table and lower the collar through the rotary. If the outside of the collar is smooth, crew members attach a safety clamp to the collar before releasing the elevators (fig. 24). The clamp holds the collar in the rotary should the slips fail. If the collar has slip recesses, the clamp is not needed. The slips are used to hold the collar in the rotary (fig. 25). Once the collar is secure, crew members are ready to repeat the process until all the collars are connected.

Figure 24. A safety clamp is attached to a smooth collar before elevators are released.

a. Drill collar slips

b. Slips holding collar in rotary

Figure 25.

Inspection

Special contractors and inspection crews inspect drill collars for damage on a regular schedule. Carefully performed field inspections can help minimize downhole failures. The frequency of inspection depends on the type of drilling being done. Experience is the best teacher for how frequently the inspection crew should inspect the drill collars. Once a month is a fairly typical inspection schedule, but the schedule can be adjusted depending on the number of cracks found or the number of failures occurring. If the inspection crew finds from one to three cracks each inspection, then once a month is frequent enough. If more than three are found, they should inspect more frequently. Crew members can also make unscheduled inspections when the collars have been used long enough to form fatigue cracks or any other damage that may be worsened downhole during drilling and cause the collar to fail.

Visual Inspections

API RP 7G, section 10, gives recommendations for drill collar inspection. The collars should be inspected to determine overall condition and obvious damage. Inspection crew members should then measure the outside and inside diameters of both ends to determine whether the outside of the collar is worn. Remember, the drilling crew must know the exact OD, ID, and style of connection (the type, size, and pitch of the threads) before looking up the recommended makeup torque in API RP 7G. The drilling crew must also know the OD, ID, and style of connection to determine if the shoulders are adequate to form a good seal between the drill collars and to make sure that the bending strength ratio between the box and the pin has not fallen below 2.50.

The inspection crew should use a profile gauge to make sure the threads on the pin have not been stretched or worn (fig. 26). Stretched or worn threads may not hold the shoulders together well enough to form a good seal during drilling.

Inspection crew members should also make sure that the box diameter is not swollen and that the box and pin shoulders are flat and free from any damage. The inspection crew can repair slight damage to shoulders by refacing and beveling shoulders. If extensive damage is found, however, the drill collar must be repaired in a machine shop.

Figure 26. A profile gauge is used to check threads on pin.

Magnetic Inspections

In addition to visually inspecting the pin and box for any signs of damage, the inspection crew should inspect the drill collars magnetically. Magnetic inspection can find cracks not easily seen. The inspection crew magnetizes the collar with an electric coil. Next, crew members spray the ends of the magnetized collars with a fluid containing fine iron filings coated with a fluorescent dye. The filings are attracted to any notch or crack. The inspection crew then uses an ultraviolet (black) light under a hood to see if any cracks or notches are present.

Remember, the inspection crew can repair slight damage to shoulders by refacing and beveling them. If extensive damage is found, the collar must be repaired in a machine shop.

To summarize—

Main functions of drill collars

- To put WOB in order to make hole
- To keep the drill string from buckling
- To provide the pendulum effect in straight-hole drilling
- To force the bit to drill a hole that is aligned with the hole already drilled

Types of drill collar design

- Standard—30- or 31-foot (9.144- or 9.44-metre) steel pipe with ends threaded in either a box or a pin configuration
- Special—square collars for extremely crooked hole drilling; spiral collars to prevent differential sticking; collars with recesses for slips and elevators

Causes of drill collar failures

- Using the improper amount of torque
- Using collars with damaged threads
- Using collars with shoulder size inadequate for a good connection
- Using inadequate amount or incorrect type of lubricant
- Using improper procedures during makeup or breakout
- Improper cleaning
- Not using protectors on collars when moving them about the rig
- Not handling the collars properly when moving them about the rig

Steps in visual inspection of drill collars

- Measure the outside and inside diameters of both ends
- Measure the OD to determine wear, and measure OD and ID to check for proper makeup torque and bending strength ratio
- Measure threads on pin to determine whether they have been stretched and, if so, how much
- Check the outside diameter of the box to make sure it is not swollen or damaged
- Reface and bevel the shoulder if only slightly damaged

*Steps in magnetic inspection of drill col*lars

- Magnetize collar with electric coil
- Spray ends with fluid containing fine iron filings coated with a fluorescent dye
- Use an ultraviolet light under a hood to see if any cracks or notches are present

▼
▼
▼

Drill String

▼
▼
▼

Unlike drill collars, the drill string is not ordinarily used to put weight on the bit. (The drill string is, however, sometimes used to put weight on the bit in horizontal drilling.) The drill string is made of steel or aluminum and is normally used for two basic purposes: to serve as a conduit, or conductor, for the drilling fluid; and to transmit the rotation of the rotary table or top drive to the bit on bottom. Since it is not exclusively used to put weight on the bit, the drill string is smaller and lighter than the drill collars. In addition, in straight-hole drilling, it is suspended in the hole under tension, not compression. It is kept in tension by two opposing forces—the weight of the collars pulling on it from below and the hoist, line, and blocks pulling on it from the surface (fig. 12). Keeping the drill string in tension prevents it from bending and buckling and prolongs its life.

Standard Design

Manufacturers design the drill string so that it can withstand some of the most complex stresses encountered during drilling. Relative to a drill collar, the drill string is small and thin, yet it can withstand powerful forces. Basically, the drill string is a column, or string, of drill pipe with attached tool joints. Most drill pipe is steel that is forged into a solid bar and then pierced to produce a seamless tube. Because the wall of the tube is relatively thin, usually less than ½ inch (12.7 millimetres) thick, the manufacturer cannot cut threads into it. To solve the problem of providing threaded ends so that the pipes can be screwed together, manufacturers produce tool joints.

Tool Joints

The tool joint is a separate piece of metal welded onto the seamless drill pipe to produce the characteristic bulge at each end (fig. 3). The wall of the tool joint is thick enough to have the pin or the box cut into it.

To prepare the drill pipe for welding, the manufacturer first heats the ends of the pipe and then strikes the heated end forcefully. These heavy end-on blows thicken the hot steel in the pipe ends. Manufacturers call the thickened ends "upsets." The pipe maker thickens the last 3 to 6 inches (76.2 to 152.4 millimetres) of each end of the pipe to make it stronger.

Manufacturers produce drill pipe with different types of upsets. Some have an internal upset; that is, the bore of the drill pipe is reduced. Viewing the pipe from the outside reveals no thickened areas. Smaller pipes have an external upset. These pipes have increased outer diameters (fig. 27). The internal diameter, or bore, of the pipe is not reduced. Most drill pipe comes with both internal and external upsets.

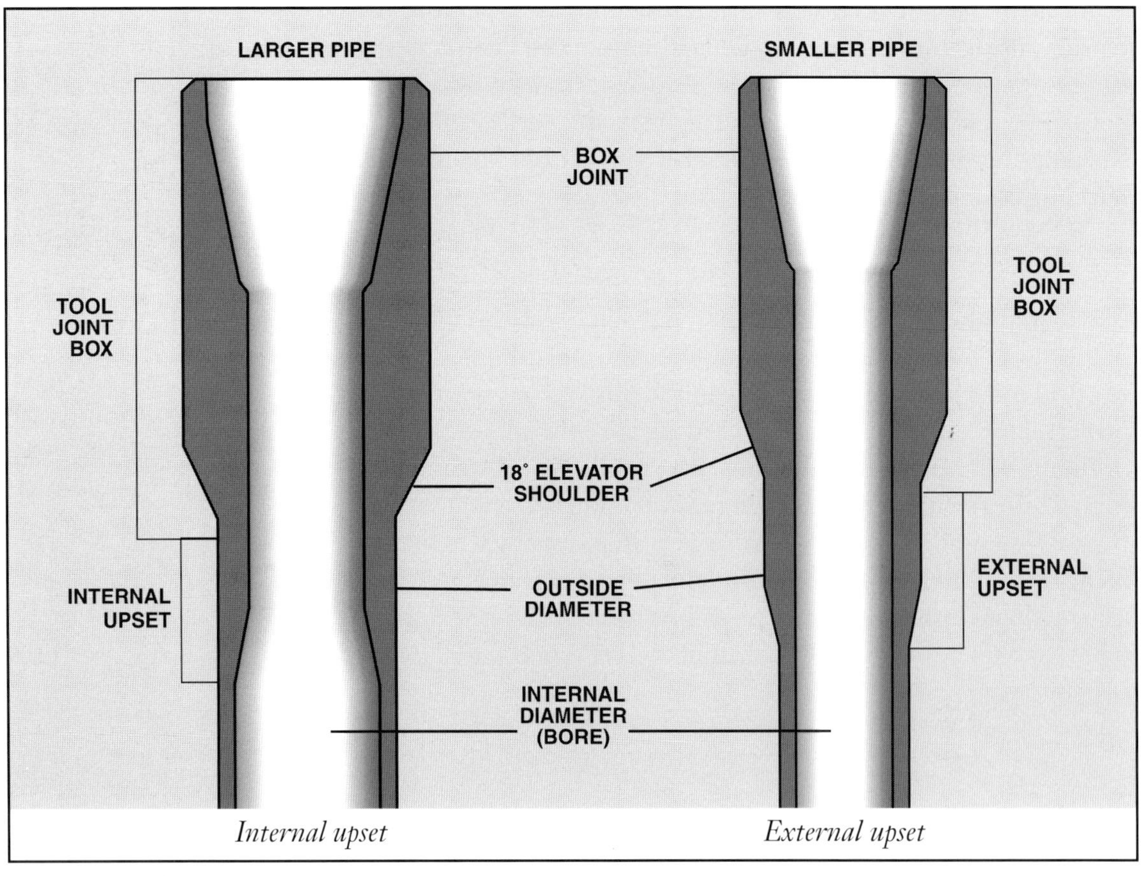

Figure 27. Pipe upset

Once the ends of the drill pipe are upset, the manufacturer welds the cylindrical tool joints to the upset ends by spinning the tool joint at a high rate of speed on a flywheel while the joint is touching the upset end of the pipe. Placing the upset end of the pipe against the spinning tool joint creates enough heat to weld the two together. This type of welding is referred to as *friction welding*, or *inertia welding*.

The wall of the tool joint is about 2 inches (50 millimetres) thick and about a foot (a third of a metre) long. Each tool joint pin and box includes the *tong area* and the elevator shoulder. The tong area refers to that part of the joint to which drilling crews attach the tongs that make up or break out the tool joints. Pipe manufacturers machine and shape the tool joint and cut threads into it to make one end of the pipe a pin and the other end a box (fig. 28).

Manufacturers produce some tool joints with hardfacing on the joint's exterior (fig. 28). When correctly applied, hardfacing may greatly increase the life of a tool joint. Tool joints tend to wear more rapidly when rotating through a dogleg or a curved portion of the hole. An abrasive formation can reduce the size of the tool joint and weaken it. The hardfacing on the tool joint can withstand abrasive wear much better than ordinary steel surfaces can and thus can prolong the life of the tool joint.

Manufacturers also produce tool joints with tapered elevator shoulders so that the pipe can more easily slide past doglegs and curves in the wellbore. Figure 28 indicates the tapered shoulders of a box and pin. Normally, the shoulder of the box has an 18° taper. Years ago API recommended that the pin shoulder should also be tapered to 18°. Producing this gradual taper on the pin, however, increased the pin length. Adding to the length of either the pin or the box increases the cost of the joint. For years, the industry agreed that only the box needed the 18° taper. Thus, API recommended a compromise taper of 35° for pin shoulders. With more horizontal drilling taking place and extended-reach wells getting longer and longer, however, the need for an 18° pin shoulder has become more apparent. Pin shoulders with this degree of taper reduce the drag on drill pipe during trips out of the hole. API has now agreed to make 18° pin shoulders an acceptable option.

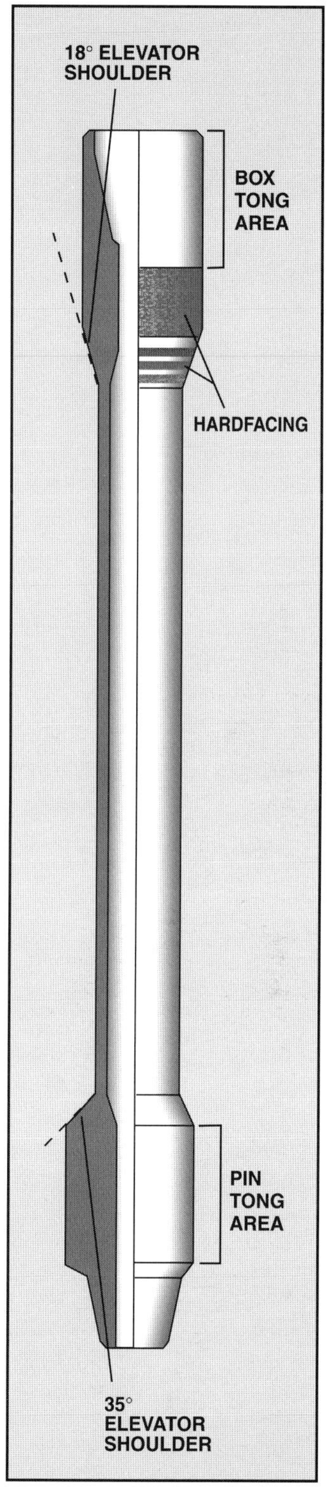

Figure 28. Drill pipe joint

API-Recommended Sizes

Each piece of drill pipe (excluding the tool joint) may have an outside diameter ranging from 2⅜ to 6⅝ inches (6.03 centimetres to 16.83 centimetres). Not only does the OD of drill pipe vary, but also the length of the pipe. Manufacturers make drill pipe in one of three API-recommended ranges of lengths. Range 1 lengths vary from 18 to 22 feet (5.49 to 6.71 metres). If a pipe measures, for example, 20 feet (6 metres) long, it would be range 1 length. Range 2 lengths fall within 27 to 30 feet (8.23 to 9.14 metres). Range 3 lengths are from 38 to 45 feet (11.58 to 13.72 metres). Manufacturers produce these three ranges of lengths because derrick heights vary. The drilling contractor must be able to buy drill pipe lengths that make into stands of a height that fit inside the derrick. For example, in a 156-foot (48-metre) derrick, the drilling crew can run three-joint stands of drill pipe (called thribbles) which total around 90 feet (27.4 metres). The drilling crew needs plenty of height above the stand to be able to manipulate it in the derrick. The contractor would, in this case, choose range 2 drill pipe, since range 2 pipe is from 27 to 30 feet (8.23 to 9.14 metres) long. The most commonly used length of drill pipe is the range 2 length. Most derricks are from 125 to 150 feet (38.1 to 45.72 metres) tall, which allows a three-joint stand of 30-foot (about 9-metre) joints to fit into the derrick.

Grades of Drill Pipe

Manufacturers produce drill pipe according to API specifications concerning *yield* and *tensile* strengths. *Minimum yield strength* refers to the specific value at which the pipe will permanently distort. *Minimum tensile strength* refers to a specific value at which the pipe will snap, or pull apart. Drilling contractors determine what type drill pipe they need based on the conditions they expect to encounter downhole. The depth of the hole is the primary factor for determining what grade drill pipe is needed. Other factors include whether or not the hole is straight or directional and the type of formations being drilled.

Most present-day seamless drill pipe falls into one of four API grades—

E-75, with a minimum yield strength of 75,000 psi (517.125 megapascals—MPa), a maximum yield strength of 105,000 psi (723.975 MPa), and a minimum tensile strength of 100,000 psi (689.5 MPa);

X-95, with a minimum yield strength of 95,000 psi (655.025 MPa), a maximum yield strength of 125,000 psi (861.875 MPa), and a minimum tensile strength of 105,000 psi (723.975 MPa);

G-105, with a minimum yield strength of 105,000 psi (723.975 MPa), a maximum yield strength of 135,000 psi (930.825 MPa), and a minimum tensile strength of 115,000 psi (792.925 MPa); and

S-135, with a minimum yield strength of 135,000 psi (930.825 MPa), a maximum yield strength of 165,000 psi (1137.675 MPa), and a minimum tensile strength of 145,000 psi (999.775 MPa).

What these grades tell the drilling crew is the force at which the drill pipe will be damaged in some way. The standard grades of pipe have a range of yield strengths (minimum to maximum) but only one minimum tensile strength. E-75 includes all drill pipe with a minimum yield strength falling somewhere between 75,000 psi (517.125 MPa) and 105,000 psi (723.975 MPa), but whose minimum tensile strength is 100,000 psi (689.5 MPa). If drill pipe has a minimum yield strength of say, 85,000 psi (586.075 MPa) and a minimum tensile strength of 100,000 psi (689.5 MPa), it would be classified as E-75 pipe. If, however, it had a minimum yield strength of 100,000 psi (689.5 MPa) with a minimum tensile strength of 105,000 psi (723.975 MPa), it would be classified as X-95. Obviously, the higher grade steel pipes have higher yield and tensile strengths and can withstand much more force than the lower grades.

Because of deeper drilling and higher stress levels, a non-standard (i.e., non-API) higher strength drill pipe has been developed—V-150 with a minimum yield and tensile strength of 150,000 psi (1034.25 MPa).

In addition to yield and tensile strength, API also provides tables reflecting the *torsional yield strength*, *burst strength*, and *collapse strength* of the four standard grades of pipe. The torsional yield strength of drill pipe is the amount of twisting force a pipe can withstand before twisting off—that is, twisting until it snaps apart. The torsional strength of the pipe itself is usually less than that of the tool joints except when a high-strength drill string is used (all grades above E-75 are considered high-strength).

The burst strength of drill string is the amount of internal pressure that a pipe can withstand before it bursts or springs a leak. The bursting pressure on a drill string in a well is the sum of the pump pressure applied at the top of the string and the hydrostatic pressure developed by the liquid inside the pipe at a particular depth. If the drill string is open at the bottom, hydrostatic pressures inside and outside the pipe are balanced at any depth.

The collapse strength of drill string is the amount of external pressure that a pipe can withstand before it collapses or caves in.

Drill String Operating Limits

Even though the drill string is manufactured according to certain tested yield and tensile strengths, it is well known that the metal in the pipe is subjected to greater stresses under dynamic conditions in the well than under the static conditions of the lab. Steel has, however, the capability of absorbing cycles of stress for an infinite number of reversals if the stress is kept under a certain limit. This limit is referred to as the *endurance* or *fatigue limit* of the steel. If the stress never exceeds that point, any number of cycles will not cause the pipe to fail. It is difficult to measure precisely the fatigue limit of drill pipe because of the various stresses that can be encountered while drilling. It is, however, presumed to be approximately one-third to one-half the minimum tensile strength of the pipe.

Drilling beyond the fatigue or endurance limit will, depending on the number of cycles of stress, eventually cause the pipe to fail. To illustrate this point, consider bending a wire coat hanger back and forth with your hands. If you bend it only slightly, you can bend it forever without breaking the wire. Repeatedly bending it back and forth to a much greater degree will cause this wire to deform and eventually to break, however.

The fatigue in any metal is progressive. Alternating stresses create heat in the section of metal that is moving back and forth and reduce the strength of the metal. Submicroscopic cracks form that eventually unite until a crack becomes visible. The crack is normally 90° to the stress. Thus, drill pipe fatigue failures are circumferential or transverse—that is, they go around the pipe (fig. 29).

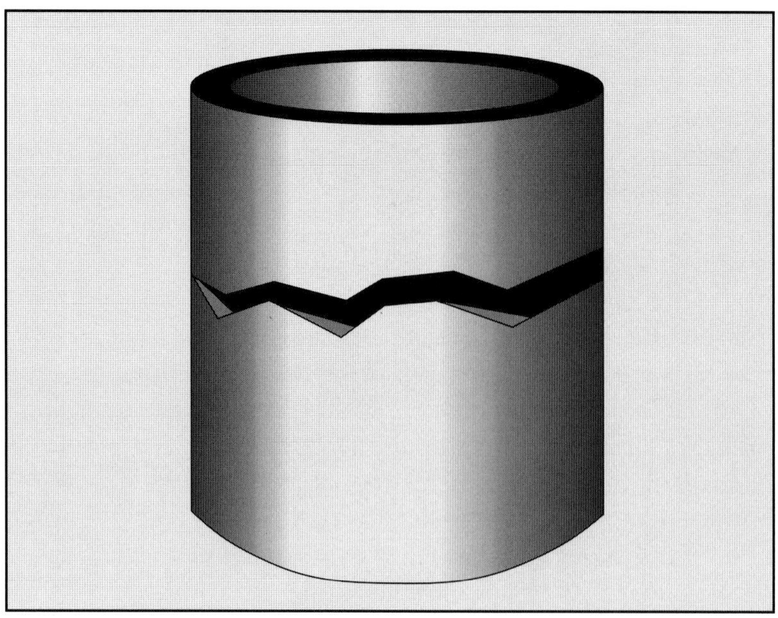

Figure 29. Drill pipe fatigue

Rotating the drill string in a straight hole normally keeps the pipe well under the fatigue limit. But drill pipe is subjected to cyclical stresses of tension, compression, and torsion, particularly in doglegs and deviated holes. In fact, the major factor in fatigue-related failures is cyclic bending, which occurs when pipe is rotated in a hole that changes direction. Bending the pipe and rotating it at the same time in a curved hole submits the pipe to alternate cycles of bending or compression and tension. With each rotation, the side of the pipe that was under tension is put under compression and the side that was under compression is under tension (fig. 30). Obviously, the greater the curve in the hole, the greater the levels of stress on the drill string.

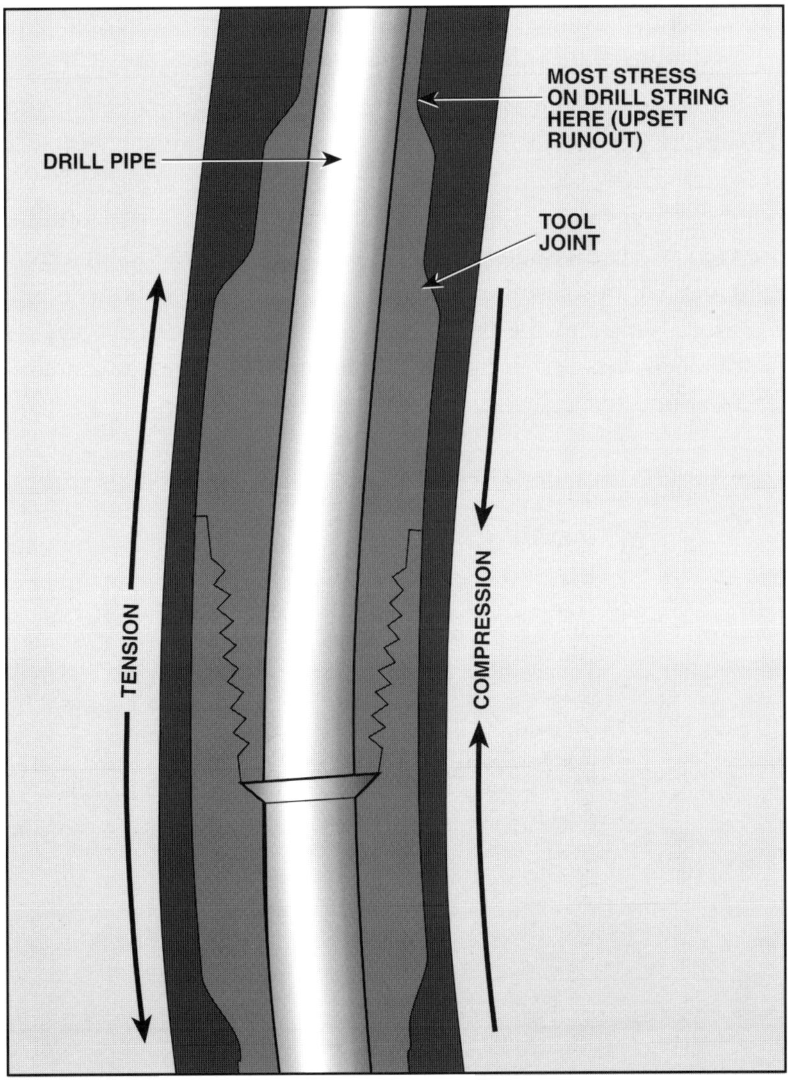

Figure 30. In a curved wellbore, more stress is put on drill pipe. The greater the curve, the greater the amount of tension and compression.

The maximum stress on the drill pipe occurs at the *upset runout*, that is, where the upset ends and the nominal wall of the pipe meet. The tool joint is strong so the most bending occurs in the relatively thin pipe wall.

The amount of fatigue damage to the pipe depends on several factors:

1. the tensile load in the pipe at the dogleg. Tensile load refers to the amount of tension, or pull, on the pipe;
2. the severity of the dogleg (usually expressed as degrees per 100 feet);
3. the number of cycles (number of rotations) in each dogleg;
4. the dimensions and properties of the pipe; and
5. the corrosibility of the environment. A very corrosive drilling fluid will drastically shorten pipe life.

If drill pipe is damaged while rotating through a dogleg, the damage is permanent. Repetitive stresses accumulate and will eventually cause failure. Drill pipe may be damaged on one hole but not fail immediately. The failure can occur later and far from the position in the drill string where the trouble started, or the drill pipe may fail in subsequent wells.

In addition to the pipe's failing as a result of metal fatigue, factors such as imperfections in the steel and bent pipe can increase the likelihood of failure.

Notch Failure

Surface imperfections in the drill pipe metal greatly affect the fatigue limit of the metal. Imperfections can be mechanical (i.e., notches or pits) or metallurgical (defects in the steel itself). A notch or pit concentrates the stresses encountered during drilling and speeds the breakdown of the metal structure (fig. 31). They are, therefore, referred to as *stress risers*, or *stress concentrators*. Where the notch or pit appears on the drill pipe determines how much it will affect the fatigue limit of the pipe. If a notch is on a portion of drill pipe not subject to stress, the notch has little effect. If a notch is within 20 inches (50.8 centimetres) of a tool joint (in the pipe's upset runout), where maximum bending takes place, it can form the nucleus of an early fatigue break.

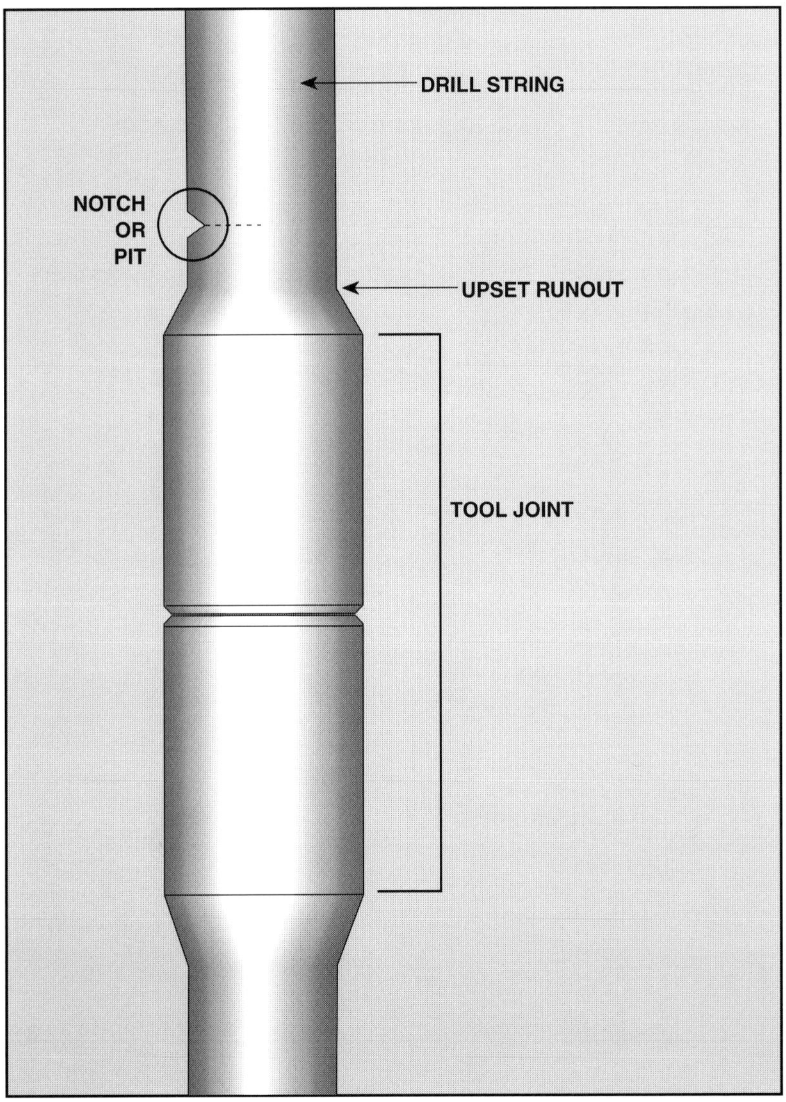

Figure 31. A notch or pit concentrates drilling stresses.

The shape and type of notch or scratch is also important. A longitudinal notch (fig. 32), an extensive saucer with a rounded bottom, will distribute the stress and be relatively harmless whereas a minute scratch with a sharp bottom will act as a stress riser and lead to failure.

Some steel is more sensitive to notches than other steel; notched brittle steel fails more quickly than ductile steel.

Various surface dents and scratches that can cause eventual drill pipe failure include the following:

1. tong marks;
2. slip marks, cuts, and scratches;
3. spinning chain marks and scratches;
4. stencil markings;
5. hammer marks;
6. corrosion grooves caused by rubber protectors;
7. electric arc burns; and
8. downhole notching caused by formation and junk cuts.

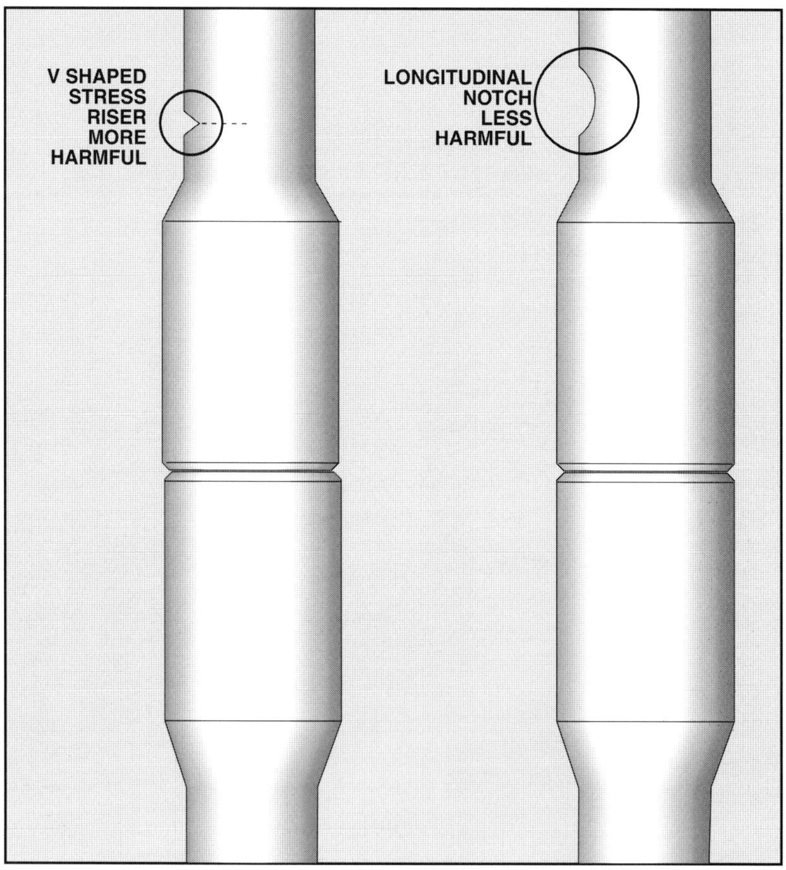

Figure 32. Two types of stress risers: notched (left) and longitudinal (right)

Of all these defects, tong marks, although rare, are probably the most damaging marks produced on drill pipe in the field. They are long, deep, and frequently sharp. Because such notches are longitudinal, they may not lead to notch failure. Even a slight deviation from vertical in the wellbore, however, can change the stress on the pipe and longitudinal notches can become stress concentrators. A change in the wellbore from vertical alters the stress along the defect from longitudinal to transverse. Tongs, therefore, should be applied to the tool joint, never to the body of the drill pipe, because the tool joint is thicker and more resistant to failure than is the pipe itself. In addition, applying tongs to the drill pipe body may crush the pipe as well as notch it.

Rotary slips are made with fine serrations (fig. 33) and are used to hold the pipe in place and to prevent it from slipping down into the hole when a connection is being made or broken out. The slips can, however, score the pipe if they are mistreated, worn, or carelessly handled. Slips with worn, mismatched, incorrectly sized, or improperly installed gripping elements (called dies) can allow one or two teeth or portions of the teeth to catch the full load of the drill string, thereby causing deep notching and potential failure.

Figure 33.

a. Drill pipe slips

b. Slips holding drill string in the hole

Inasmuch as any transverse mark can be a dangerous stress concentration point, it is not surprising that steel stencil marks can be the start of fatigue when parts of the letter are transverse to the pipe and the steel stamp is in the wrong place. Never steel stencil on the drill pipe tube.

Corrosion at the top of rubber drill pipe protectors can produce a circumferential groove. These grooves can lead to failure. Modern protectors are designed to minimize this risk. Because the protectors usually produce the grooves when they are left on the pipe while the pipe is in storage, the International Association of Drilling Contractors (IADC) recommends removing them before placing pipe in storage.

Welders sometimes attach the ground lead to the pipe rack instead of to the material being welded. This action is particularly dangerous in that the subsequent arcing between the rail and the drill pipe goes unnoticed. This arcing pits the pipe. Though these pits are small, they are surrounded by a wide band of burned metal that is as hard as glass and very prone to rapid fatigue failure.

Bent Pipe

It is very important that the drilling crew not run bent or crooked pipe into the hole. A crooked joint of pipe is always a potential failure. A crooked kelly can cause bending in the first joint of drill pipe below the rotary table. If the stress is great enough, failure will occur. Having a crown block off center because the mast or derrick is not plumb can also cause pipe failure because the off-center block throws bending stresses into the kelly and the drill string.

Corrosion Fatigue

Corrosion is the alteration and degradation of material caused by its environment. Corrosion fatigue, or metal failure due to a corrosive environment, is a common cause of drill stem failures. With water-based drilling fluid, the chief corrosive agents of drill pipe are dissolved gases (oxygen, carbon dioxide, and hydrogen sulfide), dissolved salts, and acids.

Most modern drill pipe is made with a thermally baked plastic coating applied to the inner surface to minimize corrosion pitting. Wirelines and tools in the drill string bore tend to rupture or destroy the plastic coating that protect the pipe.

Among the many factors affecting corrosion rates of drill pipe are the following:

1. *pH.* pH is a scale for measuring the hydrogen ion concentration of a particular environment. The pH scale is logarithmic; that is, each pH increment of 1.0 represents a tenfold change in hydrogen ion concentration. The pH of pure water is 7.0. pH values below 7 are increasingly acidic, and pH values greater than 7 are increasingly alkaline. In the presence of dissolved oxygen, the corrosion rate of steel in water is relatively constant between pH 4.5 and 9.5. The corrosion rate increases rapidly at lower pH values and decreases slowly at higher pH values. In drilling, the pH level rarely falls below 7. Most problems occur at pH levels between 7 and 10.5.
2. *Temperature.* In general, corrosion rates increase with increasing temperature.
3. *Velocity.* In general, corrosion rates increase with higher rates of fluid flow through the pipe.
4. *Heterogeneity.* In general, the more uniform the grain structure of a pipe, the less will be the corrosive effect of the environment. Localized variation in composition or microstructure of the metal—that is, corrosion in small, well-defined areas—may increase the corrosion rates.
5. *High stresses.* Highly stressed areas may corrode faster than areas of lower stress. The highest bending stresses occur in doglegs, where the tension is highest.

Corrosion can take many forms and may combine with other destructive processes—erosion, abrasive wear, and notch failures—to cause severe damage. Several forms of corrosion may occur at the same time, but one type will usually predominate. The following forms of corrosion are most often encountered with drill pipe:

1. *Uniform, or general, attack.* In this type of corrosion, the pipe corrodes evenly, usually leaving a coating of corrosion products (e.g., iron oxide, or rust).
2. *Pitting, or localized, attack.* In this type of corrosion, the pipe corrodes in small, well-defined areas, causing pits to form. These pits may vary in number, depth, and size; they may also be obscured by corrosion products. A drill stem inspection crew can detect pitting with magnetic inspection (see section on magnetic inspection for details). Pits can serve as points of origin for fatigue cracks and lead to washouts. A *washout* is a place where a small opening has occurred in the pipe, usually as a result of a fatigue crack's penetrating the pipe wall and allowing drilling fluid through it.
3. *Corrosion fatigue.* In a corrosive environment, no fatigue limit exists, since failure will ultimately occur from corrosion, even in the absence of cyclic stress. The cumulative effect of corrosion and cyclic stress is greater than the sum of the damage from each. The endurance limit, or fatigue threshold, will always be lower in a corrosive environment, even under mildly corrosive conditions that show little or no visible evidence of corrosion.

Sulfide Stress Cracking

When drilling through H_2S environments, sulfide stress cracking (SSC)—a form of hydrogen embrittlement—is a frequent cause of drill stem failure (see appendix figs. 56, 57). Both stress and the absorption of hydrogen in the presence of hydrogen sulfide are involved in this type of failure.

Atomic hydrogen, the smallest of atoms, is a product of most corrosion reactions. It can be absorbed by and diffused through steel and other metals. Normally, the hydrogen atoms quickly combine to form molecular hydrogen, which is too large to be absorbed by the metal and bubbles off as a gas. In the presence of sulfide, however, the hydrogen remains in the atomic form for a considerably longer time and therefore has a greater probability of being absorbed by the pipe. After being absorbed, the hydrogen tends to accumulate in the area of maximum stress and, when a critical concentration is reached, a small crack forms. The hydrogen accumulates at the top of the crack and the crack grows. This process continues until the remaining metal cannot sustain the applied load and a sudden brittle fracture occurs.

The degree of this effect on a piece of steel is determined by the concentration of hydrogen, the strength of the steel, applied stress on the steel, and the time of exposure:

1. *Strength of the steel.* Generally, the higher the strength (grade) of the steel, the greater is its susceptibility to SSC. The lowest strength drill pipe capable of withstanding the required drilling conditions should be used.
2. *Total tensile load (stress) on the steel.* The higher the total tensile load on the pipe, the greater is the possibility of failure by SSC. Each grade of steel has a critical, or threshold, stress below which SSC will not occur; however, the higher the grade, the lower the threshold stress.
3. *Amount of atomic hydrogen and H_2S.* The higher the amount of atomic hydrogen and H_2S present, the shorter the time before failure.
4. *Time.* Time is required for atomic hydrogen to be absorbed and diffused in the steel to the critical concentration required for a crack to begin and failure to occur. By controlling the factors in the previous three listings, the time before which failure occurs may be sufficiently lengthened to permit the use of marginally susceptible steels for a short duration .
5. *Temperature.* The severity of SSC is greatest in normal atmospheric temperatures; it decreases as temperature increases. Operating at temperatures in excess of 135°F (57°C) allows marginally susceptible materials to be used in potentially embrittling environments. The greater the hardness of the material, the higher the required safe operating temperature. Drillers must be careful, however, because SSC failure may occur when the material returns to normal temperature after it is removed from the hole.

To minimize the risk of SSC in water-based drilling fluids, drillers should—

1. control the drilling fluid pH. When practical, given other functions of the drilling fluid, the driller should maintain a pH of 10 or higher. In drill strings containing aluminum drill pipe, the pH should not exceed 10.5 because aluminum pipe tends to corrode more than steel at a high pH level.
2. limit gas-cutting and formation fluid invasion of the wellbore by maintaining proper drilling fluid weight. Hydrogen sulfide invades the wellbore from the formations being drilled.
3. chemically treat the drilling fluid for H_2S inflows from formations, preferably prior to encountering the sulfide.
4. use the lowest-strength drill pipe capable of withstanding the required drilling conditions.
5. use care in tripping out of the hole after exposure to an H_2S environment and avoid sudden shocks and high loads.
6. remove absorbed hydrogen from the pipe after exposure to an H_2S environment by aging the pipe in open air for several days to several weeks, depending on the exposure conditions, or bake it at 400° to 600°F (204°C to 316°C) for several hours. Plastic-coated drill pipe should not be heated above 400°F (204°C).
7. limit drill stem testing in H_2S environments to as brief a period as possible, using operating procedures (using H_2S inhibitors, for example) that will minimize exposure to SSC conditions.

Corrosion and SSC can be minimized by the use of oil-based drilling muds. Corrosion does not occur if metal is completely enveloped by an oil environment that is electrically nonconductive. Therefore, under drilling conditions that cause serious problems of corrosion damage, erosion-corrosion, or corrosion fatigue, drill stem life can be greatly extended by using an oil mud.

Care and Handling of the Drill String

Because the drill string is such an expensive part of the rig and its components, it must be handled with care. The tool joints, in particular, are susceptible to damage and yet have to hold up to extreme pressures, rotations, torque, bending, and other downhole stresses of drilling.

Thread Protectors

Thread protectors will prevent most of the tool joint damage that occurs in moving and racking the pipe. A thread protector is a device that is screwed into the box or onto the pin of a tool joint to keep the threads and shoulders from being damaged while the pipe is moved or stored. The protectors fit over the pin and box and are available in pressed steel, cast steel, or plastic (fig. 34). When not in use, the protectors can be screwed together to protect their threads from damage that could later damage mating tool joint threads.

a. Pin thread protectors

b. Box thread protectors

Figure 34.

Cleaning

Some drilling crews thoroughly clean the pin and box threads and shoulders before making up tool joints. Cleaning them pays off in three ways:

1. It removes foreign material and allows for proper makeup with other tool joints. Proper makeup helps the tool joint perform in a trouble-free manner.
2. It permits better inspection of the threads and shoulders.
3. It increases the life of the joints by eliminating abrasive materials.

Drilling crews should clean new boxes and pins with a suitable solvent and a soft bristle brush because a wire brush might remove the antigalling treatment applied by the manufacturer. The crew should make sure the threaded joints are completely clean and dry before inspecting and applying thread compound. Solvents can dilute the thread compound used during makeup and increase the chances of metal-to-metal contact and galled threads.

Inspection

After cleaning or before making up one tool joint to another, the crew should carefully inspect the threads and shoulders of each joint. A damaged joint should never be made up to another joint and run into the hole. Even slight damage can cause improper makeup of the two joints and will likely cause wobbling or leaking connections. Some damage can be repaired in the field—slight crowning (high spots) on the shoulders, slight cups (low spots), dents or upsets, fins (sharp ridges around the box or pin shoulder), and minor galls. More serious damage requires repair in a machine shop.

All shoulders not obviously in need of repair should be checked for flatness with a test ring. Before using the ring, a crew member should make sure the shoulder is clean and dry. The crew member holds the ring against the shoulder by applying pressure with the fingers at two diametrically opposed points (fig. 35) and tries to make the ring rock. The crew member then repeats the procedure at points 90° from the first points of pressure. If the ring rocks at all, the shoulder is either rough or crowned and should be faced off flat, or smoothed until flat, with a shoulder dressing tool.

Figure 35. Checking the tool joint shoulder for flatness

The crew should take great care when facing, or dressing, a shoulder. It is easy to take off too much of the shoulder. It is good practice to remove no more than 1⁄32 inch (.80 millimetre) at any one facing and never more than 1⁄16 inch (2⁄32 inch, or 1.6 millimetres) cumulatively on each shoulder. The crew should use the box or pin benchmark to gauge the total amount of facing (fig. 36). The benchmark for the pin is on the base of the pin—1⁄8 inch (3.20 millimetres) from the shoulder. The benchmark for the box is on the inside of the box—1⁄8 inch (3.2 millimetres) from the shoulder. To make measuring easier, the crew may want to convert all fractions to the same denominator—the 1⁄8 inch measurement to 4⁄32 inch (3.20 millimetres) and the 1⁄16 inch to 2⁄32 inch (1.6 millimetres), for example.

After facing a pin shoulder for the first time, the distance from the benchmark to the pin shoulder should be no more than 5⁄32 inch (4.0 millimetres). The cumulative distance for all pin refacings should be no more than 6⁄32 inch (4.8 millimetres).

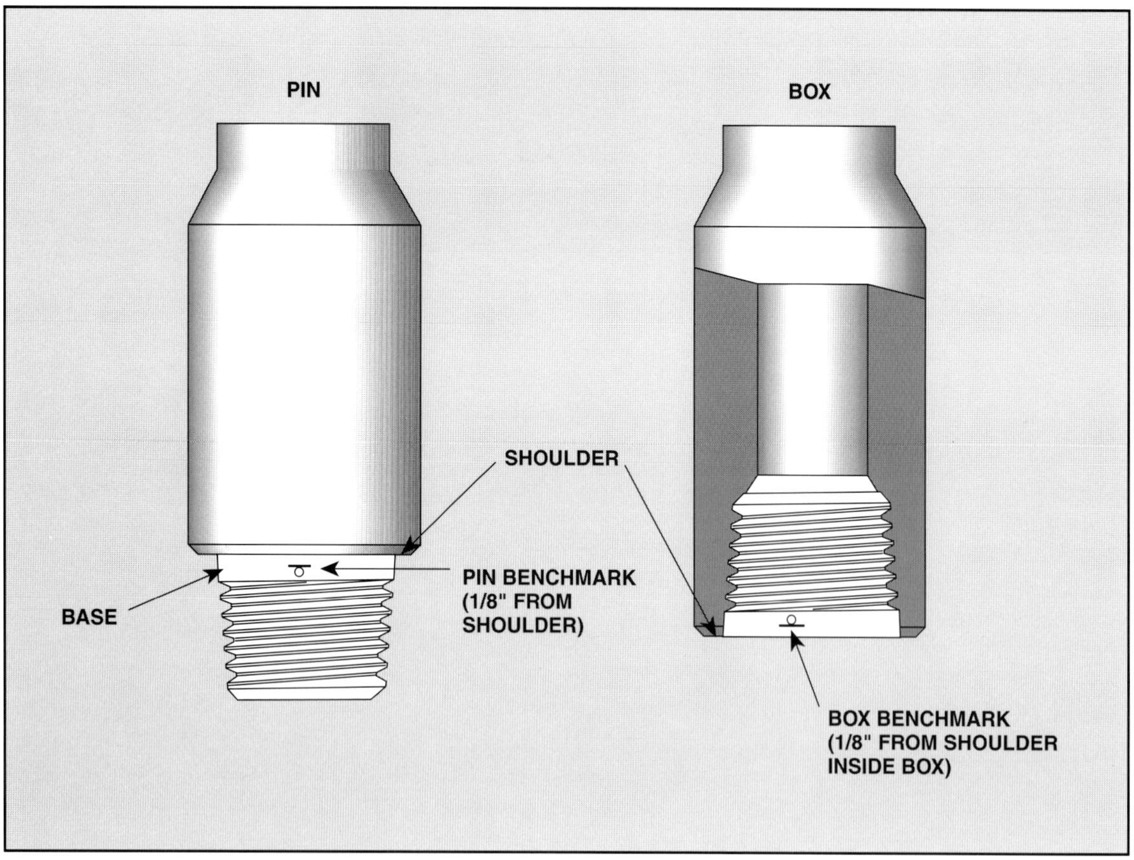

Figure 36. Pin and box benchmarks

After facing a box shoulder for the first time, the distance from the shoulder to the benchmark should be no less than 3⁄32 inch (2.4 millimetres). The cumulative total for all box refacings should be no less than 2⁄32 inch (1.6 millimetres) cumulatively.

The box's outside diameter and the pin's inside diameter should be measured in order to determine the proper amount of torque required during makeup. The crew should measure the OD at a distance of 1 inch (25 millimetres) from the box shoulder. Once crew members know the OD, they look in the correct API table to determine the proper torque for this particular tool joint.

An independent inspection crew should check the tool joint threads with a thread profile gauge to make sure they have not been stretched, worn, or damaged in any way not easily seen (fig. 26). If the inspection crew finds evidence of pin stretching, it should perform a magnetic particle inspection of the entire threaded area of the pin, especially the last engaged pin thread (fig. 5), to determine if cracks are present. (For a description of how this inspection is performed, see "Magnetic Inspections" under Drill Collars). In addition, the inspection crew checks the OD of the drill pipe tube, performs an end-area magnetic test, and performs an electronic magnetic inspection (EMI) test of the entire tube length to check for cracks.

Lubricating

When making up joints of drill string, the crew should apply a good coating of the proper thread compound, called dope or lubricant, to the tool joint threads and shoulders. This lubricant, normally a soft metallic component like copper or zinc in a grease base, prevents metal-to-metal contact and protects the threads from galling. It also makes it easier to screw the two joints together. The crew should never thin or dilute the compound lubricant because this will reduce the percentage of metallic component and thus increase the chances of galling the threads and shoulders. Some cleaning fluids will dilute the compound and keep it from adhering properly to the surfaces it is meant to protect. For this reason, the crew should make sure the pin and box threads and shoulders are completely clean and dry after cleaning with a solvent and before applying the thread compound.

IADC and API recommend compounds containing from 40% to 60% by weight of finely powdered metallic zinc for drill pipe and drill collar connections. Thread compounds, such as API-modified, are made especially for tubing and casing and should not be used on tool joints. These lubricants are slick and will allow the joint to be overtightened with a low torque application. Tool joints can be severely damaged if tubing or casing lubricants are used.

API-Recommended Torque

Just as using the proper amount of torque is important when connecting drill collars, using the correct amount of torque is also very important when connecting the drill string. Inadequate torque during makeup can cause leaks, wobbles, washouts, and galls. Too much torque applied during makeup can also cause problems, including swelled or split tool joint boxes and stretched pins (fig. 37). API recommends proper makeup torque for both drill collars and tool joints. Tables for these values are available in API RP 7G. This information is also available in the IADC *Drilling Manual*, 11th ed., section B. This section of the IADC manual also details procedures for handling new pipe, tripping in and out of the hole, and moving pipe around the rig. All of these procedures are designed to protect the pipe and keep it in good condition.

Figure 37. Swelled tool joint box

Making a Good Connection

In order to reduce the number of drill string failures, crew members should make up tool joints correctly—that is, they should make a *good connection*. They should use the proper torque (the proper amount of turning force needed to screw the tool joints together as determined by API RP 7G), proper procedures, and adequate lubricant, or dope. In a good connection, the shoulders of the tool joints support the pin so that it is rigidly held by the box. In a good connection, the seal between the two joints is made between the shoulders of the pin and the box. The pin and the box serve as a simple screw jack to pull the two shoulders together very tightly. Unlike the threads in tubing and casing, where the threads produce a continuous seal, the threads of the tool joints do not form a pressure seal. In a good connection, the shoulders on the tool joints must be connected with adequate *preload*, or proper makeup torque, to produce a seal that will hold even under the tension and compression put on the tool joints during drilling.

Making a Bad Connection

A crew can make a bad connection by using too little torque, too much torque, an inadequate amount of lubricant, or improper procedures. Any one of these practices can damage the tool joint.

When crew members use too little torque, tension in drilling can cause an opening between the shoulders of the pin and the box. Once this opening occurs, the tool joint connection is said to wobble. A loose joint, or wobble, breaks the seal between the shoulders. Every time the seal breaks, drilling fluid passes across the shoulder (fig. 38).

If drilling fluid passes across the shoulder, three things happen: (1) the lubricant, or dope, between the shoulders washes out; (2) the surface of the shoulders begins to erode; and (3) if enough fluid moves through the seal, the lubricant is washed away from the threads as well as from the shoulders. The crew can recognize a loose joint when breaking out drill pipe because the pin is dry (no lubricant is present).

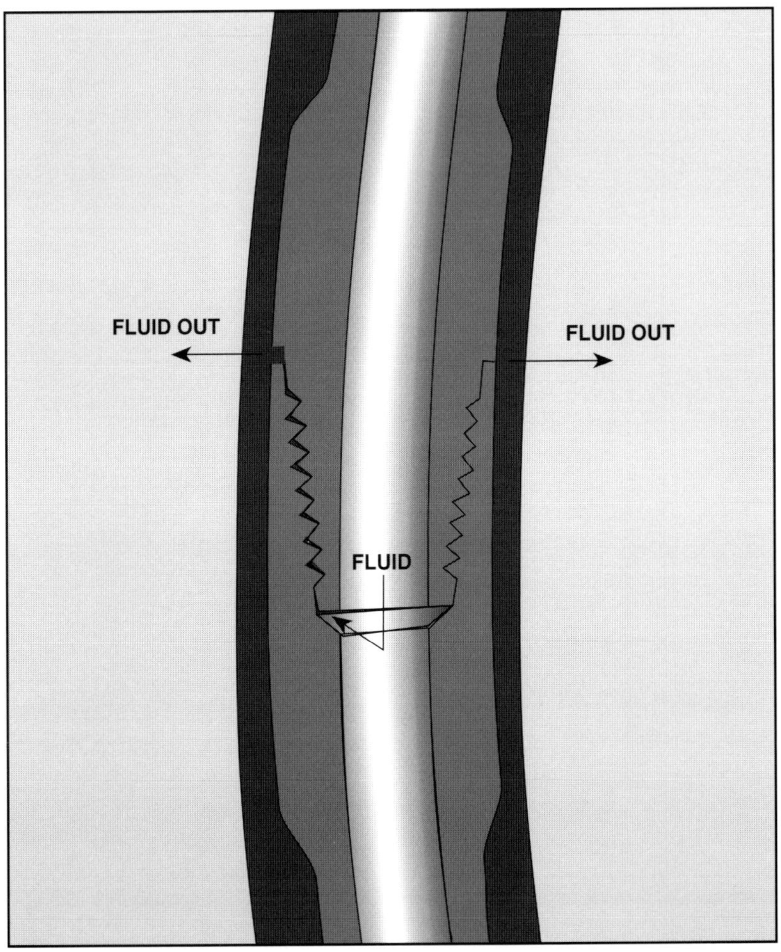

Figure 38. A loose joint can allow drilling fluid to leak across threads and shoulders.

Drilling fluid coming out through the shoulders causes threads to erode in a wobble. If this situation goes unchecked, it can result in a complete washing away of the threads and the beginning erosion of the body of the box itself.

High stress loads on mating threads and shoulders and inadequate or improperly applied thread compounds may lead to galling. Galling occurs when two mating steel surfaces, not protected by a film of lubricant (such as metallic zinc or copper) freeze together. When the connection is broken out, the metal is torn apart and the threads and shoulders are damaged (fig. 19). Galling at its worst can destroy one or both makeup shoulders and can wipe out all the threads in a pin or box. It occurs frequently on newly machined surfaces.

Too much torque on the tool joint can cause the pin to stretch or the box to bell. Stretching the pin too far can break it (fig. 39), resulting in a lost drill string and an expensive fishing job.

Figure 39. Broken pin

Bad stabbing practices can also lead to drill string failure. When a joint has been cleaned, checked, and lubricated, it is ready to be made up to another tool joint and lowered into the hole. The crew must be sure that one joint does not strike against the rig or any other piece of equipment. Pounding or dropping the tool joint will damage it. When stabbing the pin into the box, the crew should do so carefully so the pin does not hit the box shoulder.

The crew should make sure that the shoulders of the box and the pin match. Using different-width shoulders can produce *fins*, which are thin, sharp ridges around the box or pin shoulders (fig. 40).

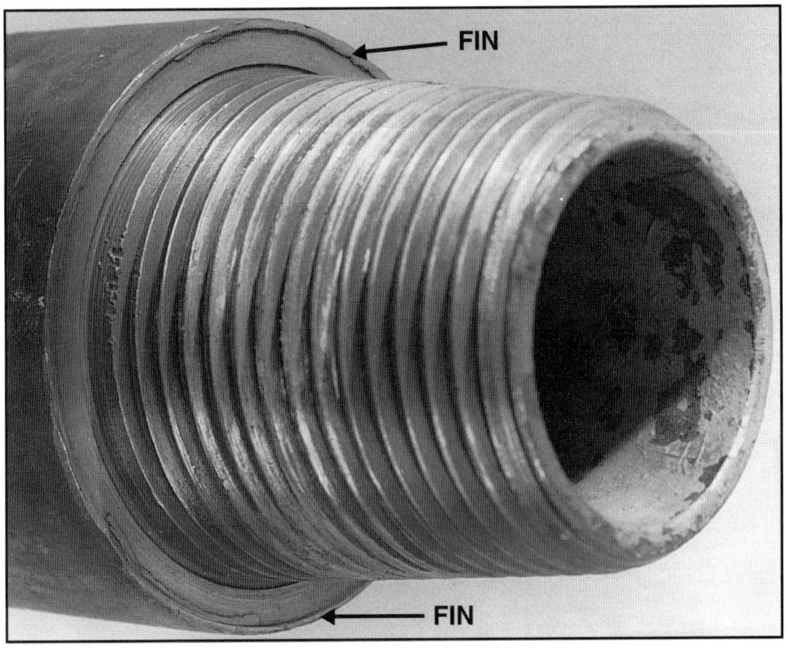

Figure 40. Fin on pin shoulder

To make sure that the shoulder and the tool joint threads match for a connection, the crew should measure them with a gauge called a tool joint identifier. This tool measures the number of threads per inch and diameter of the rotary shouldered connection (fig. 41).

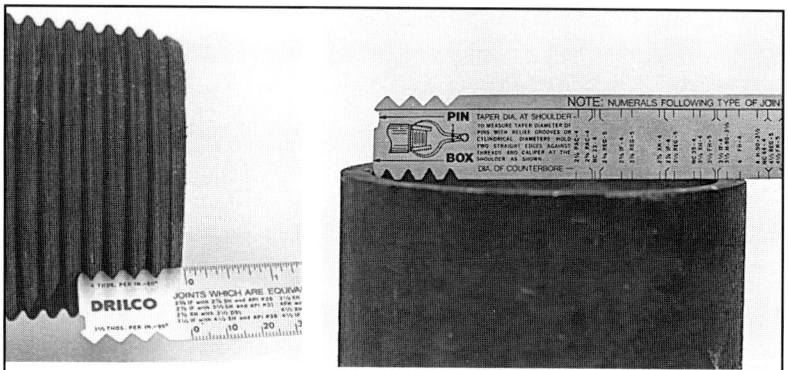

a. Measuring the number of threads per inch

b. Measuring the diameter of the bore

Figure 41.

To summarize—

Main functions of the drill string

- To serve as a conduit, or conductor, for the drilling fluid
- To transmit the rotation of the rotary or top drive to the bit on bottom

Components of the drill string

- A pierced, seamless tube of forged steel or extruded aluminum
- Tool joints attached to each end of the seamless tube

Characteristics of tool joints

- Separate pieces of metal attached to the seamless tube to produce a characteristic bulge at each end of the drill pipe
- Thick enough to have a pin or box cut into them
- Provide the connections for drill string

Manufacturing specifications of drill pipe

- API-recommended sizes so that the industry has standard sizes
- API-recommended grades, depending on the specific yield and tensile strengths required so that drilling contractors can determine the right type of pipe for the specific drilling situation

Operating limits of drill string

- Can withstand infinite cycles of stress if the stress is kept under the endurance or fatigue limit of the steel
- Drill string failure if operating beyond the fatigue limit
- Several factors increase the stress on the drill string: rotating drill string in curved holes; corrosive environments; hydrogen sulfide; notches, pits, scratches, tong marks, electric arc burns, etc.

Care and handling of drill string

- Inspect the threads and shoulders of each joint to make sure damaged tool joints are not joined together
- Check the threads with a thread profile gauge to make sure they have not been stretched, worn, or damaged in any way not easily seen

- Perform a magnetic particle inspection of the entire pin and box threaded areas to determine if cracks are present
- Check the drill pipe tube with an electronic magnetic inspection to determine if cracks are present
- Measure the outside diameter of the tool joint to determine the proper amount of torque required during makeup
- Apply a good coating of thread compound to tool joint threads and shoulders
- Use the API-recommended amount of torque when making up tool joints
- Use the proper procedures when picking up and laying down drill string, or making up or breaking out drill string so that it is not jarred, scratched, or damaged in any way.

Heavy-Walled Drill Pipe

▼
▼
▼

Design

Heavy-walled (heavy-weight) drill pipe is manufactured with walls that are thicker than those in standard drill pipe. The heavier-walled tube is attached to special extra-length tool joints. The extra length of the tool box allows room for recutting connections when the original ones are damaged and reduces the rate of wear on the OD of the tube by keeping the wall of the tube away from the side of the hole. The OD of the tube is also protected from abrasive wear with a center wear pad (fig. 42).

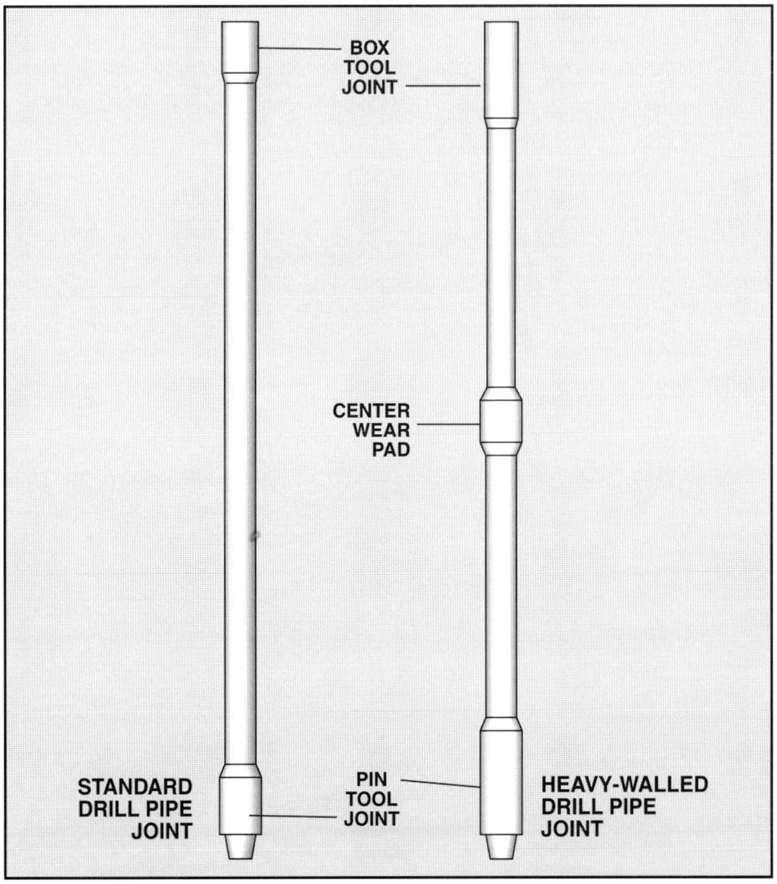

Figure 42. Standard drill pipe joint and heavy-walled drill pipe joint with center wear pad

Some drilling calls for a specially designed heavy-walled drill pipe. This heavy pipe is spiraled—much like the spiral drill collar. Figure 43 compares the spiral heavy-walled pipe and the conventional heavy-walled pipe. Both have extra-length tool joints but the spiral pipe does not have a center wear pad.

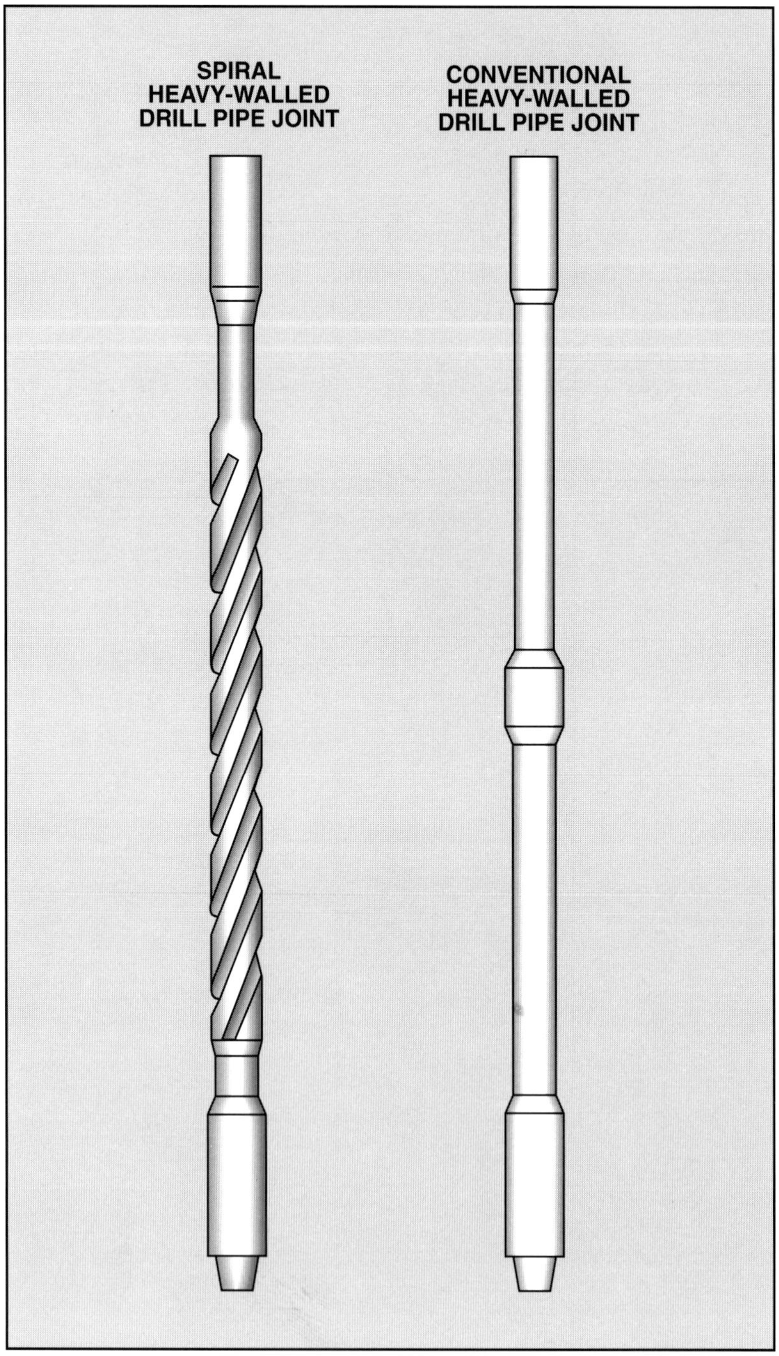

Figure 43. Two types of heavy-walled joints

Function

Heavy-walled pipe is frequently used in the drill stem just above the drill collars, in the transition zone between the stiffer collars and the more limber drill string, and in place of some drill collars to apply weight on the drill bit in small-diameter holes.

In Transition Zones

Fatigue failures often occur in the first few joints above the drill collars. This area of the drill string is referred to as the *transition zone*. The rigid, heavy collars do not bend as much as the more limber drill string. Most of the bending stress is placed on the first few joints of drill string immediately above the drill collars. To reduce the number of failures, drillers continually move the first joint of drill string from this high-stress location to another spot higher in the drill string. Moving these joints to other locations in the string does not correct the cumulative fatigue damage, however; it only postpones the time until fatigue failure occurs. Using five to seven joints of heavy-walled drill pipe above the collars will provide a graduated change in stiffness in this section of the drill stem, thus greatly reducing the fatigue damage in this transition zone.

In Directional Drilling

Drillers often use heavy-walled pipe in conjunction with fewer drill collars in high-angle drilling. It is less rigid than drill collars and has much less hole or wall contact. Its three-point wall contact (fig. 44) helps solve two serious problems in directional wells—drill stem wear and tear and differential sticking. Heavy-walled pipe can also be run through hole angle and direction changes with fewer connection and fatigue problems than with collars.

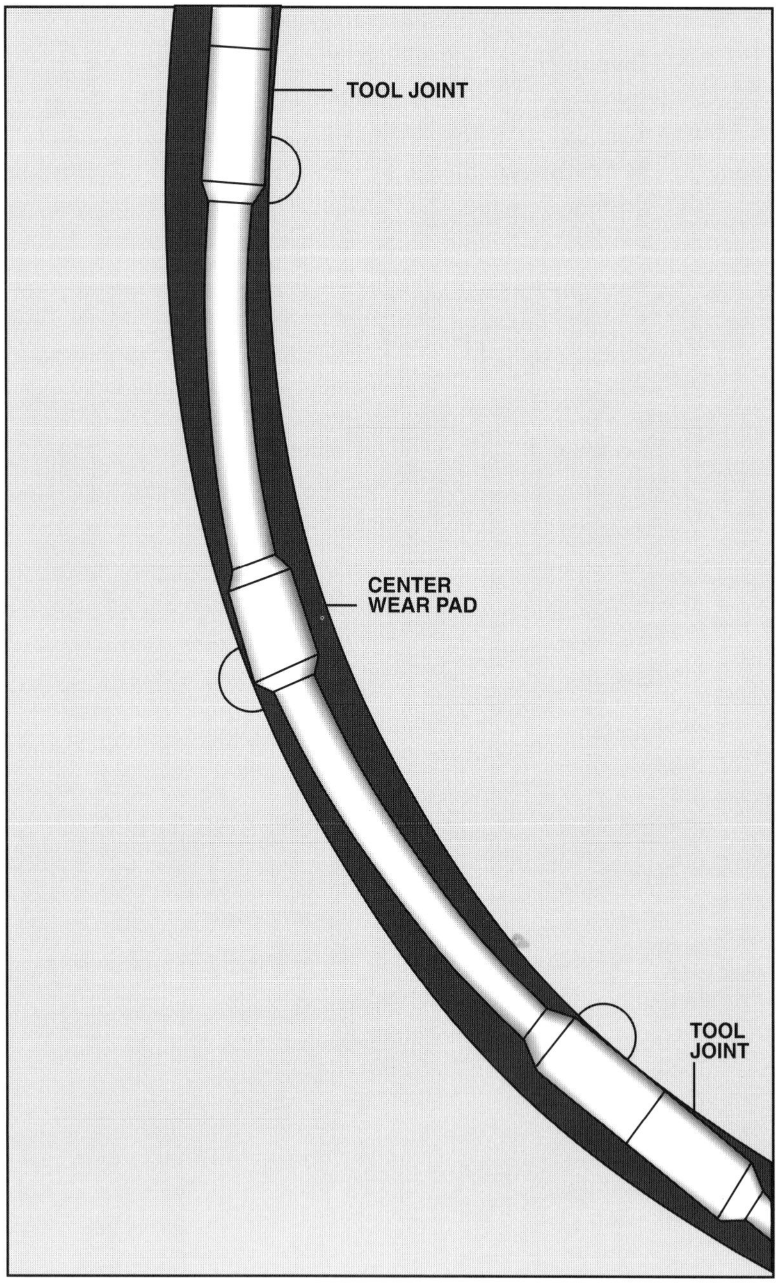

Figure 44. Three-point wall contact allows heavy-walled pipe to help solve two problems—drill stem wear and tear and differential sticking.

Heavy-walled pipe is used—

- in high-angle drilling
- in compressive service—a type of drilling in which the drill stem configuration downhole is in the following order: the bit, heavy-walled drill pipe, and collars. In compression drilling, the heavy-walled drill pipe is not held in tension but rather is being compressed by the weight of the drill collars above it
- in the transition zone between conventional collars and standard drill pipe.

For more information on directional drilling and drilling under compression, see Rotary Drilling Series, Unit III, Nonroutine Operations, Lesson 1: *Controlled Directional Drilling.*

In Straight-hole Drilling in Soft Formations

Drilling crews use heavy-walled drill pipe or spiral-shaped heavy-walled pipe to replace some of the collars when drilling vertically in soft formation. Using these heavier pipes solves some problems that using collars cannot. Using heavy-walled pipe—

1. allows the crew to reduce torque in high-rpm deep drilling by reducing the amount of contact between the outside diameter of the pipe and the wall of the hole.
2. reduces the amount of time it takes to trip in and out of the hole by eliminating the need to attach lifting subs to the collars. The elevators can be directly attached to the heavy-walled pipe.
3. eases the handling and transportation duties from one location to another because the pipe is lighter and easier to maneuver than collars are.

To summarize—

Design of heavy-walled drill pipe—

- Heavier walls and longer tool joints than standard drill pipe, and a center wear pad
- Heavier walls and longer tool joints than standard drill string, and spirals instead of a center wear pad

Functions of heavy-walled drill pipe—

- Used in the transition zone between the collars and the drill string to prevent drill string buckling
- Used in directional drilling and compression drilling
- Used in place of drill collars to prevent differential sticking

▼
▼
▼

Drill String and Drill Collar Auxiliaries

▼
▼
▼

A drilling crew installs additional pieces of equipment, or tools, in the drill string when required. These tools include subs, pup joints, stabilizers, reamers, and special valves. A *sub*, which is short for substitute, is a fitting crew members insert into the string to perform a special function. One commonly used sub is a short fitting with different-sized threads on each end. It allows the crew to screw together components of the drill string that have different-sized or -style threads.

For instance, the crew sometimes uses a *bit sub* between the bit and the drill collar. Bits come with a pin rather than a box; the pin screws into a box. The crew runs drill pipe and drill collars into the hole with the pins pointing down and the boxes facing up to make it easier for the rotary helpers to stab pins into boxes. The crew therefore cannot directly connect the bit to the drill collar, because two pins face each other. A bit sub has a box connection at each end (fig. 45). The crew stabs the drill collar's pin into the box of the bit sub, then the bit's pin into the sub's box. (Many contractors use a special bottomhole drill collar that has a box connection on each end. The box connection on the bottom of the collar allows the crew to directly stab the bit's pin into the collar without having to use a sub.)

Rig crews often use a *crossover sub* to connect the last joint of the drill string to the first drill collar in the drill stem, since the drill string's pin threads usually do not fit properly into the drill collar's box threads. Using the sub helps facilitate a good connection between the drill pipe string and the drill collars.

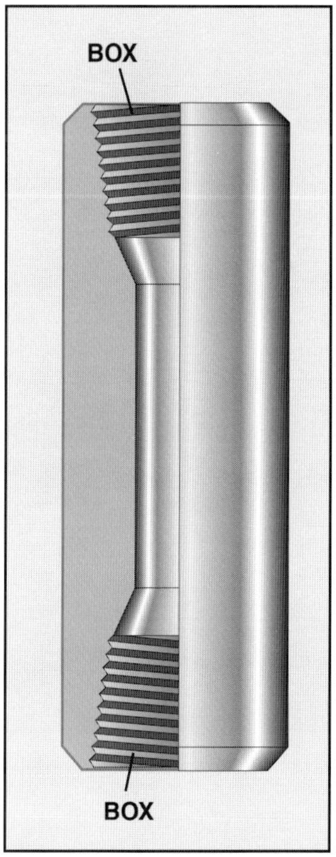

Figure 45. A bit sub; drawing shows a quarter-section taken out to show box threads inside the sub.

Another common sub is a *lifting sub* (fig. 23). The crew makes it up on the end of a drill collar. A lifting sub gives a drill collar an elevator shoulder like the ones found on drill string tool joints. Whether on a tool joint or a lifting sub, a shoulder provides a place to latch elevators. The elevators hang from the traveling block. They come into play when the crew sets the kelly back to trip the drill stem in or out of the hole. When the crew latches the elevators onto the shoulder of a tool joint, the traveling block can lift the drill stem. A lifting sub provides a place for the crew to latch the elevators onto the drill collar.

A *pup joint* is a length of drill pipe or casing that is shorter than range 1. Any joint of pipe shorter than 18 feet, or 5.5 metres, is therefore a pup joint. Crew members may need to make up a pup joint when a special procedure requires them to suspend the drill stem to an exact depth in the hole. For example, assume that they have to test a formation that is at a particular depth. To position the test tool at the exact depth, the crew may need a pup joint. Sometimes even a range 1 joint is too long.

The crew places *stabilizers* at various points in the drill collar string to keep the hole straight or to intentionally curve it from vertical (fig. 46). Blades on the stabilizer contact the wall of the hole as the drill string rotates. This contact helps keep the hole on course. In many cases, the crew tries to drill the hole as straight as it can. In other cases, however, the crew deviates the hole from vertical. Offshore, for example, a platform is very small when compared to the size of a reservoir, so the crew drills many directional, or curved, wells that tap different areas of the reservoir. Stabilizers help keep the hole straight or help deviate it, depending on where the crew places them in the string.

a. Rotary helpers make up a stabilizer in the drill collar string.

b. Stabilizers

Figure 46. Stabilizers

In some cases, hard, abrasive formations grind down the sides of a bit to a size smaller than its manufactured diameter. Drillers say such a bit is "undergauge." Undergauge bits drill undergauge hole—that is, hole smaller in diameter than the manufactured size of the bit. Full-gauge bits (bits whose diameters are full size) and other full-gauge tools can get stuck in undergauge hole. To minimize undergauge hole, the crew sometimes runs reamers or reamer-stabilizers (fig. 47). Special cutters on the reamers or reamer-stabilizers contact the formation and enlarge the hole to full size, or gauge.

Figure 47. A reamer-stabilizer

To summarize—

Additional equipment used with the drill string and drill collars

- Subs—used as connectors between different drill string, or drill string and drill collar, or bit and drill collar
- Pup joints—a length of the drill string that is shorter than range 1 and used to suspend the drill string to an exact depth in the hole
- Stabilizers—used to keep the hole straight or to intentionally curve the hole from vertical
- Reamers—used to enlarge the hole to full gauge

▼
▼
▼

Appendix

Quick Guide to Problem Solving

▼
▼
▼

Figure 48
Problem: Undertorquing wobble
Effect: Downhole pin failure
Possible repercussions: String separation, fishing, lost time
Possible solutions: Check API torque tables; repair joint, if possible.

Figure 49
Problem: Improper torque or lubrication
Effect: Swelled box
Possible repercussions: Distorted connection, loss of seal, improper mating with pin
Possible solutions: Check API torque tables; recut box, if possible; use recommended tool joint lubricant; use thicker tool joint.

Figure 50
Problem: Undertorquing
Effect: Severe pin erosion
Possible repercussions: Reduced mud pressure, leaking seals, lost string
Possible solutions: Check API torque tables; recut or reface connection, if possible; keep thread protectors on as much as possible; clean and lubricate shoulders before makeup; use only pipe jack with wide area of contact.

Figure 51
Problem: Mismatched shoulders
Effect: Fins
Possible repercussions: Erosion, washout, leaking seal, connection wobble, bad shoulder connection
Possible solutions: Remove fins by beveling; handle pipe with thread protectors.

Figure 52
Problem: Undertorquing
Effect: Erosion
Possible repercussions: Reduced mud pressure, leaking seals, lost string, bad connection
Possible solutions: Check API torque tables; recut or reface connection, if possible; clean and lubricate shoulders before makeup.

Figure 53
Problem: Stress
Effect: Fatique crack and washout
Possible repercussions: String separation, pipe loss, need for fishing
Possible solutions: Avoid critical rotary speeds; use vibration dampeners; dampen stress; minimize rate of hole deviation; use lowest-strength pipe possible; prevent H_2S in flow.

Figure 54
Problem: Dropped slips
Effect: Slip cuts where edge of slip opening caught OD of pipe
Possible repercussions: Stress concentration, surface notching, fatigue cracks, lost string
Possible solutions: Use backup tongs; set slips carefully; improve slip and bowl maintenance.

Figure 55
Problem: H_2S, water, or salt corrosion
Effect: Corrosion pits
Possible repercussions: Body-wall loss, surface notching, stress concentration, washout
Possible solutions: Dampen stress; keep mud pH above 9.5; use plastic coating, inhibitors, and oxygen scavengers; clean pipe ID and OD; use corrosion test rings.

Figure 56
Problem: Cyclic stressing, hydrogen embrittlement, holes in pipe
Effect: Fatigue cracking and necking down
Possible repercussions: Pipe loss, washout or twistoff, string separation
Possible solutions: Minimize surface notching; avoid critical rotary speeds; bring bottomhole pipe up hole on trips; use shock subs; dampen stress; avoid H_2S in flow; choose lowest-strength drill pipe possible; minimize rate of hole deviation.

Figure 57
Problem: H_2S, water, or salt corrosion
Effect: Corrosion pits
Possible repercussions: Body-wall loss, surface notching, stress concentration, fatigue cracks
Possible solutions: Dampen stress; keep mud pH above 9.5; use plastic coating, inhibitors, and oxygen scavengers; clean pipe ID and OD; use corrosion test rings.

Figure 58

Problem: Cyclic stressing, hydrogen embrittlement, holes in pipe

Effect: Fatigue cracking

Possible repercussions: Pipe loss, washout or twistoff, string separation

Possible solutions: Minimize surface notching; avoid critical rotary speeds; bring bottomhole pipe up hole on trips; use shock subs; dampen stress; avoid H_2S in flow; choose lowest-strength drill pipe possible; minimize rate of hole deviation.

Figure 59

Problem: Stress bending downhole

Effect: Fatigue cracking

Possible repercussions: Pipe loss, washout or twistoff, string separation

Possible solutions: Minimize surface notching; avoid critical rotary speeds; bring bottomhole pipe up hole on trips; use shock subs; dampen stress; avoid H_2S in flow; choose lowest-strength drill pipe possible; minimize rate of hole deviation.

Figure 60

Problem: Ring corrosion

Effect: Corrosion and pitting

Possible repercussions: Stress concentration, grooving and notching of tubular goods, fatigue cracks

Possible solutions: Periodically remove protector and clean pipe, including area under protector.

Glossary

▼
▼
▼

A

abrasion *n*: wearing away by friction.

acid brittleness *n*: see *hydrogen embrittlement*.

acidity *n*: the quality of being acid. Relative acid strength of a liquid is measured by pH. A liquid with a pH below 7 is acid. See *pH*.

air weight *n*: the weight of an object in air as opposed to the weight of an object suspended or floating in liquid. Compare *buoyant weight*.

American Petroleum Institute (API) *n*: oil trade organization (founded in 1920) that is the leading standardizing organization for oilfield drilling and producing equipment. It maintains departments of transportation, refining, and marketing in Washington, DC. Its offers publication regarding standards, recommended practices, and bulletins. Address: 1220 L St., NW; Washington, DC 20005; (202) 682-8000.

annular *adj*: pertaining to the annulus. The annulus is sometimes referred to as the annular space.

annular space *n*: the space between two concentric circles. In the petroleum industry, it is usually the space surrounding a pipe in the wellbore; sometimes termed the annulus.

API *abbr*: American Petroleum Institute.

API-certified *adj*: said of a tool that meets the American Petroleum Institute's minimum standards.

auxiliaries *n pl*: equipment on a drilling or workover rig that is not a direct part of the rig's drilling equipment, such as the equipment used to generate electricity for rig lighting or the equipment used to mix drilling fluid.

B

backup tongs *n pl*: the tongs used to back up the drill pipe as it is being made up into or taken out of the drill stem. See *breakout tongs, makeup tongs*.

backup wrench *n*: any wrench used to hold a pipe or a bolt to prevent its turning while another length of pipe or a nut is being screwed into or out of it.

BHA *abbr*: bottomhole assembly

bit *n*: the cutting or boring element used in drilling oil and gas wells. The bit consists of a cutting element and a circulating element. The cutting element is steel teeth, tungsten carbide buttons, industrial diamonds, or polycrystalline diamond compacts (PDCs). These teeth, buttons, or diamonds penetrate and gouge or scrape the formation to remove it. The circulating element permits the passage of drilling fluid and utilizes the hydraulic force of the fluid stream to improve drilling rates. In rotary drilling, several drill collars are joined to the bottom end of the drill pipe column, and the bit is attached to the end of the drill collars. Drill collars provide weight on the bit to keep it in firm contact with the bottom of the hole. Most bits used in rotary drilling are roller cone bits, but diamond bits are also used extensively.

bit breaker *n*: a heavy plate that fits in the rotary table and holds the drill bit while it is being made up in or broken out of the drill stem. See *bit*.

bit drift *n*: the tendency of the bit to move other than vertically, caused by an interaction between the rotation of the bit and the varying resistance of the formation being drilled.

bit sub *n*: a sub inserted between the drill collar and the bit. See *sub*.

bore *n*: 1. the inside diameter of a pipe or a drilled hole. 2. the diameter of the cylinder of an engine. *v*: to penetrate or pierce with a rotary tool.

borehole *n*: a hole made by drilling or boring; a wellbore.

bottomhole assembly *n*: the portion of the drilling assembly below the drill pipe. It can be very simple—composed of only the bit and drill collars—or it can be very complex and made up of several drilling tools.

box *n*: the female section of a connection. See *tool joint*.

box and pin *n*: see *tool joint*.

box threads *n pl*: threads on the female section, or box, of a tool joint. See *tool joint*.

break out *v*: 1. to unscrew one section of pipe from another section, especially drill pipe while it is being withdrawn from the wellbore. During this operation, the tongs are used to start the unscrewing operation. 2. to separate, as gas from a liquid or water from an emulsion.

breakout cathead *n*: a device attached to the catshaft of the drawworks that is used as a power source for unscrewing drill pipe; usually located opposite the driller's side of the drawworks. See *cathead*.

breakout tongs *n pl*: tongs that are used to start unscrewing one section of pipe from another section, especially drill pipe coming out of the hole. See *lead tongs*, *tongs*.

brittleness *n*: the state of having rigidity but little tensile strength. Compare *toughness*.

buckling stress *n*: bending of pipe that may occur because of hole deviation. Pipe may bend because of the angle of the hole or because of an abrupt deviation such as a dogleg.

buck up *v*: to tighten up a threaded connection (such as two joints of drill pipe).

buoyancy *n*: the apparent loss of weight of an object immersed in a fluid. If the object is floating, the immersed portion displaces a volume of fluid the weight of which is equal to the weight of the object.

buoyant effect *n*: see *buoyancy*.

buoyant weight *n*: the weight of the drill stem in a mud-filled borehole. Buoyant weight is less than the weight of the drill stem in air because of the buoyant effect of the mud on the drill stem.

burst pressure *n*: the amount of internal pressure, or stress, required to burst casing or other pipe. When the pipe's internal pressure is greater than its external pressure, the pipe bursts.

burst rating *n*: the pressure at which a manufacturer has determined that a pipe or vessel will burst from internal pressure.

burst strength *n*: a pipe or vessel's ability to withstand rupture from internal pressure.

C

cake *n*: see *filter cake*.

cake consistency *n*: the character or state of the drilling mud filter cake. According to API RP 13B, such notations as "hard," "soft," "tough," "rubbery," and "firm" may be used to convey some idea of cake consistency.

cake thickness *n*: the thickness of drilling mud filter cake.

cathead *n*: 1. a spool-shaped attachment on the end of the catshaft, around which rope for hoisting and moving heavy equipment on or near the rig floor is wound. 2. an automatic cathead. See *breakout cathead, makeup cathead.*

chain tongs *n pl*: a hand tool consisting of a handle and chain that resembles the chain on a bicycle. In general, chain tongs are used for turning pipe or fittings of a diameter larger than that which a pipe wrench would fit. The chain is looped and tightened around the pipe or fitting, and the handle is used to turn the tool so that the pipe or fitting can be tightened or loosened.

chase threads *v*: to clean and deburr the threads of a pipe so that it will make up properly.

circulate *v*: to pass from one point throughout a system and back to the starting point. For example, drilling fluid is circulated out of the suction pit, down the drill pipe and drill collars, out the bit, up the annulus, and back to the pits while drilling proceeds.

circulating components *n pl*: the equipment included in the drilling fluid circulating system of a rotary rig. Basically, the components consist of the mud pump, rotary hose, swivel, drill stem, bit, and mud return line.

cold-work *v*: to work metal without the use of heat. Compare *fire bending*.

collapse pressure *n*: the amount of force needed to crush the sides of pipe until it caves in on itself. The pipe collapses when the pressure outside it is greater than the pressure inside it.

collapse resistance *n*: the ability of the wall of a pipe or vessel to resist collapse.

collapse strength *n*: the amount of external pressure that a pipe can withstand before it collapses or caves in.

collar *n*: 1. a coupling device used to join two lengths of pipe. A combination collar has left-hand threads in one end and right-hand threads in the other. 2. a drill collar. See *drill collar*.

collar pipe *n*: heavy pipe used between the drill pipe and the bit in the drill stem. See *drill collar*.

come out of the hole *v*: to pull the drill stem out of the wellbore to change the bit, to change from a core barrel to the bit, to run electric logs, to prepare for a drill stem test, to run casing, and so on. Also called trip out.

compression drilling *n*: the practice of using the drill stem to push the bit through the formations, usually done in high angle or horizontal drilling. In normal drilling, the drill stem is kept in tension by the drawworks and drilling line.

connection *n*: 1. a section of pipe or fitting used to join pipe to pipe or to a vessel. 2. a place in electrical circuits where wires join. 3. the action of adding a joint of pipe to the drill stem as drilling progresses.

controlled directional drilling *n*: see *directional drilling*.

corrosion *n*: any of a variety of complex chemical or electrochemical processes, e.g., rust, by which metal is destroyed through reaction with its environment.

corrosion fatigue *n*: metal fatigue concentrated in corrosion pits. See *fatigue*.

corrosion inhibitor *n*: a chemical substance that minimizes or prevents corrosion in metal equipment.

corrosion test *n*: one of a number of tests to determine qualitatively or quantitatively the corrosion-inducing compounds in a product.

corrosiveness *n*: the tendency to wear away a metal by chemical attack.

corrosive product *n*: a hydrocarbon product that contains corrosion-inducing compounds in excess of the specification limits for a sweet product.

coupling *n*: 1. in piping, a metal collar with internal threads used to join two sections of threaded pipe. 2. in power transmission, a connection extending longitudinally between a driving shaft and a driven shaft. Most such couplings are flexible and compensate for minor misalignment of the two shafts.

crew *n*: 1. the workers on a drilling or workover rig, including the driller, derrickman, and rotary helpers. 2. any group of oilfield workers.

critical weight *n*: weight placed on the bit that results in tension on the drill string, which causes the drill string to vibrate at the rotary speed being used. A drill stem operating with critical weight and at the critical speed for that weight will have stresses develop that cause very rapid failure.

crooked hole country *n*: an area in which particular subsurface formations make it difficult to keep a drilled hole straight.

crossover sub *n*: a sub that allows different sizes and types of drill pipe to be joined.

cross-thread *v*: to screw together two threaded pieces when the threads of the pieces have not been aligned properly.

crown *n*: 1. the crown block or top of a derrick or mast. 2. the top of a piston. 3. a high spot formed on a tool joint shoulder as the result of wobble.

cup *n*: a low spot formed on a tool joint shoulder as the result of a wobble or as a result of the tool joint being connected to another damaged tool joint. See *wobble*.

cyclic stressing *n*: stress that occurs on a pipe, vessel, or machine in cycles, such as the sucker rod string.

D

deviation *n*: departure of the wellbore from the vertical, measured by the horizontal distance from the rotary table to the target. The amount of deviation is a function of the drift angle and hole depth. The term is sometimes used to indicate the angle from which a bit has deviated from the vertical during drilling.

diameter *n*: the distance across a circle, measured through its center. In the measurement of pipe diameters, the inside diameter is that of the interior circle and the outside diameter that of the exterior circle.

directional drilling *n*: 1. intentional deviation of a wellbore from the vertical. Although wellbores are normally drilled vertically, it is sometimes necessary or advantageous to drill at an angle from the vertical. Controlled directional drilling makes it possible to reach subsurface areas laterally remote from the point where the bit enters the earth. It often involves the use of deflection tools. 2. a technique of river crossing in pipeline construction in which the pipe is buried under the riverbed at depths much greater than those of conventional crossings. With this technique, a hole in the form of an inverted arc is drilled beneath the river, and the made-up pipeline is pulled through it.

directional hole *n*: a wellbore intentionally drilled at an angle from the vertical. See *directional drilling*.

dogleg *n*: 1. an abrupt change in direction in the wellbore, frequently resulting in the formation of a keyseat. 2. a sharp bend permanently put in an object such as a pipe.

dope *n*: a lubricant for the threads of oilfield tubular goods. Also called thread compound. *v*: to apply thread lubricant.

double *n*: a length of drill pipe, casing, or tubing consisting of two joints screwed together. Compare *fourble, single, thribble.*

double board *n*: the name used for the working platform of the derrickhand (the monkeyboard) when it is located at a height in the derrick or mast equal to two lengths of pipe joined together. Compare *fourble board, thribble board.*

downhole *adj, adv*: pertaining to the wellbore.

drawworks *n*: the hoisting mechanism on a drilling rig. It is essentially a large winch that spools off or takes in the drilling line and thus raises or lowers the drill stem and bit.

dressing a shoulder *v*: in drilling, using a shoulder dressing tool to cut or grind the shoulder of a collar or tool joint until it is smooth and flat. Also called facing, or refacing.

drift angle *n*: the angle at which a wellbore deviates from the vertical, expressed in degrees, as revealed by a directional survey. Also called angle of deviation, angle of drift, and inclination.

drill *v*: to bore a hole in the earth, usually to find and remove subsurface formation fluids such as oil and gas.

drill bit *n*: the cutting or boring element used for drilling. See *bit*.

drill collar *n*: a heavy, thick-walled tube, usually steel, placed between the drill pipe and the bit in the drill stem. Several drill collars are used to provide weight on the bit and to provide a pendulum effect to the drill stem. When manufactured to API specifications, a drill collar joint is 30 or 31 feet (9.14 or 9.45 metres) long.

The outside diameter of drill collars made to API specifications ranges from 3.5 inches to 11 inches (7.9 to 27.9 centimetres).

drill collar sub *n*: a sub made up between the drill string and the drill collars that is used to ensure that the drill pipe and collar can be joined properly.

drill column *n*: see *drill stem*.

driller *n*: the employee directly in charge of a drilling or workover rig and crew. The driller's main duty is operation of the drilling and hoisting equipment, but this crew member is also responsible for downhole condition of the well, operation of downhole tools, and pipe measurements.

drill floor *n*: also called rig floor or derrick floor. See *rig floor*.

drilling crew *n*: a driller, a derrickhand, and two or more helpers who operate a drilling or workover rig for one tour each day.

drilling fluid *n*: circulating fluid, one function of which is to lift cuttings out of the wellbore and to the surface. Other functions are to cool the bit and to counteract downhole formation pressure. Although a mixture of barite, clay, water, and chemical additives is the most common drilling fluid, wells can also be drilled by using air, gas, water, or oil-base mud as the drilling mud. See *mud*.

drilling line *n*: a wire rope used to support the drilling tools. Also called the rotary line.

drilling mud *n*: a specially compounded liquid circulated through the wellbore during rotary drilling operations. See *drilling fluid, mud*.

drilling rate *n*: the speed with which the bit drills the formation; usually called the rate of penetration.

drilling rig *n*: see *rig*.

drill pipe *n*: seamless steel or aluminum pipe made up in the drill stem between the kelly or top drive on the surface and the drill collars on the bottom. During drilling, it is usually rotated while drilling fluid is circulated through it. Drill pipe joints are available in three ranges of length: 18 to 22 feet (5.5 to 6.7 metres), 27 to 30 feet (8.2 to 9.1 metres), and 38 to 45 feet (11.6 to 13.7 metres). The most popular length is 27 to 30 feet (8.2 to 9.1 metres). It is available with outside diameters ranging from 2.9 to 5.5 inches (7.3 to 14 centimetres). Several joints are made up (screwed together) to form the drill string.

drill pipe protector *n*: an antifriction device of rubber or steel attached to each joint of drill pipe to minimize wear.

drill pipe rubber *n*: a rubber or elastomer disk that is placed around a joint of drill pipe and is held stationary below the rotary table. As pipe is removed from the well, the rubber wipes mud off the outside of the pipe to minimize corrosion.

drill pipe safety valve *n*: a special valve used to close off the drill pipe to prevent backflow during a kick. It has threads to match the drill pipe in use.

drill pipe slips *n pl*: see *slips*.

drill stem *n*: all members in the assembly used for rotary drilling from the swivel to the bit, including the kelly, drill pipe and tool joints, drill collars, stabilizers, and various specialty items. Compare *drill string*.

drill string *n*: the column, or string, of drill pipe with attached tool joints that transmits fluid and rotational power from the kelly to the drill collars and bit. Often, especially in the oil patch, the term is loosely applied to both drill pipe and drill collars. Compare *drill stem*.

drill string float *n*: a check valve in the drill string that will allow fluid to be pumped into the well but will prevent flow from entering the string.

ductility *n*: see *toughness*. Compare *brittleness*.

dynamic loading *n*: exerting force with continuous movement, i.e., cyclic stressing.

E

effective weight on bit *n*: in drilling, the weight on the bit (as produced by the drill collars) minus the buoyant effect that the drilling mud has on the drill collars.

elevators *n pl*: clamps that grip a stand of casing, tubing, drill pipe, or sucker rods so that the stand can be raised from or lowered into the hole.

endurance limit *n*: see *fatigue limit*.

environment *n*: 1. the sum of the physical, chemical, and biological factors that surround an organism. 2. the water, air, and land and the interrelationship that exists among and between water, air, and land and all living things. 3. as defined by the US government, the navigable waters, the waters of the contiguous zone, the ocean waters, and any other surface water, groundwater, drinking water supply, land surface, subsurface strata, or ambient air within the United States.

equilibrium *n*: a state of balance between opposing forces or actions that is either static or dynamic.

external upset end (EUE) *n*: on tubing, casing, or drill pipe, the thickening at each end of the joint such that the internal diameter of the joint is not affected; i.e., it remains uniform throughout the joint's length. Only the outside diameter is enlarged at each end. Pipe is thickened, or upset, at each end to increase its strength so that threads, couplings, or tool joints may be attached. Compare *internal upset end*, *internal-external upset end*.

F

facing a shoulder *v*: in drilling, using a shoulder dressing tool to smooth and flatten the shoulder of a collar or tool joint. Also called *dressing* a shoulder, or *refacing* a shoulder

fatigue *n*: the tendency of material such as a metal to break under repeated cyclic loading at a stress considerably less than the tensile strength shown in a static test.

fatigue crack *n*: in drilling, a crack formed in steel as a result of repeated cyclic loading at a stress considerably less than the tensile strength shown in a static test.

fatigue limit *n*: in drilling, the value, or point, below which steel will not fail even with repeated cycles of stress

fatigue threshold *n*: see *fatigue limit*.

female connection *n*: a pipe, coupling, or tool threaded on the inside so that only a male connection can be joined to it.

filter cake *n*: 1. compacted solid or semisolid material remaining on a filter after pressure filtration of mud with a standard filter press. Thickness of the cake is reported in thirty-seconds of an inch or in millimetres. 2. the layer of concentrated solids from the drilling mud or cement slurry that forms on the walls of the borehole opposite permeable formations; also called wall cake or mud cake.

filter cake thickness *n*: a measurement of the solids deposited on filter paper in thirty-seconds of an inch during a standard 30-minute API filter test. See *cake thickness*. In certain areas the filter cake thickness is a measurement of the solids deposited on filter paper for 7.5 minutes.

fin *n*: a thin, sharp ridge around the box or the pin shoulder of a tool joint, caused by the use of boxes and pins with different-sized shoulders.

fire bending *n*: one of the earliest methods for bending pipe. The joint was first placed over a small bonfire and, when the heat had rendered it sufficiently malleable, it was placed against a tree and pressure was applied until the desired bend was achieved. Fire bends significantly weakened the pipe. A cold-work process is less damaging.

fish *n*: an object that is left in the wellbore during drilling or workover operations and that must be recovered before work can proceed. It can be anything from a piece of scrap metal to a part of the drill stem. *v*: 1. to recover from a well any equipment left there during drilling operations, such as a lost bit or drill collar or part of the drill string. 2. to remove from an older well certain pieces of equipment (such as packers, liners, or screen liner) to allow reconditioning of the well.

fitting *n*: a small, often standardized, part (such as a coupling, valve, or gauge) installed in a larger apparatus.

flash welding *n*: 1. a form of resistance butt welding used to weld wide, thick members or members with irregular faces together, and tubing to tubing. 2. in pipeline construction, a welding technique in which low voltage is applied to each pipe joint while the ends are in light contact. This contact produces a rapid arcing, called flashing. After the pipe ends have been adequately heated, the current is abruptly increased, and the pipe joints are brought together rapidly and forcefully. The current is then reduced, excess flash material in the pipe is cleared, and the weld is completed.

fluid *n*: a substance that flows and yields to any force tending to change its shape. Liquids and gases are fluids.

formation *n*: a bed or deposit composed throughout of substantially the same kind of rock; often a lithologic unit. Each formation is given a name, frequently as a result of the study of the formation outcrop at the surface and sometimes based on fossils found in the formation.

fourble *n*: a section of drill pipe, casing, or tubing consisting of four joints screwed together. Compare *double, single, thribble*.

fourble board *n*: the name used for the working platform of the derrickman, or the monkeyboard, when it is located at a height in the derrick equal to approximately four lengths of pipe joined together. Compare *double board, thribble board*.

friction *n*: resistance to movement created when two surfaces are in contact. When friction is present, movement between the surfaces produces heat.

friction welding *n*: in drilling, the proces of adhering the tool joint to the drill pipe by placing the tool joint on the upset end of the drill pipe while it is spinning, thereby creating enough heat to weld the two together.

G

gall *n*: damage to steel surfaces caused by friction and improper lubrication.

gauge *n*: 1. the diameter of a bit or the hole drilled by the bit. 2. a device (such as a pressure gauge) used to measure some physical property. *v*: to measure size, volume, depth, or other measurable property.

go in the hole *v*: to lower the drill stem, tubing, casing, or sucker rods into the wellbore.

good connection *n*: the process of correctly connecting two tool joints by using the API-recommended amount of torque, the proper procedures, and an adequate amount of thread compound, or lubricant.

gun-barrel approach *n*: a method of drilling using a packed bottomhole assembly. In this method, the new hole drilled is in line with that which has already been drilled so that the wellbore is as straight as the barrel of a gun.

H

H_2S *form*: hydrogen sulfide.

hard banding *n*: a special wear-resistant material often applied to tool joints to prevent abrasive wear to the area when the pipe is being rotated downhole.

hardfacing *n*: an extremely hard material, usually crushed tungsten carbide, that is applied to the outside surfaces of tool joints, drill collars, stabilizers, and other rotary drilling tools to minimize wear when they are in contact with the wall of the hole.

heat a connection *v*: to loosen a collar or other threaded connection by striking it with a hammer. Also called warm a connection or whip a connection.

heavyweight drill pipe *n*: drill pipe having thicker walls and longer tool joints than usual and an integral wear pad in the middle. Several joints of this pipe may be placed in the drill stem between drill collars and regular drill pipe to reduce the chances of drill pipe fatigue or failure.

hoist *n*: 1. an arrangement of pulleys and wire rope or chain used for lifting heavy objects; a winch or similar device. 2. the drawworks. *v*: to raise or lift.

hole *n*: in drilling operations, the wellbore or borehole. See *borehole, wellbore.*

hole angle *n*: the angle at which a hole deviates from vertical.

hole drift *n*: the amount a wellbore is deflected from vertical.

horizontal drilling *n*: deviation of the borehole at least 80° from vertical so that the borehole penetrates a productive formation in a manner parallel to the formation. A single horizontal hole can effectively drain a reservoir and eliminate the need for several vertical boreholes.

hydrogen embrittlement *n*: low ductility of a metal caused by its absorption of hydrogen gas. Also called acid brittleness.

hydrogen ion concentration *n*: a measure of the acidity or alkalinity of a solution, normally expressed as pH. See *pH.*

hydrogen sulfide *n*: a flammable, colorless gaseous compound of hydrogen and sulfur (H_2S), which in small amounts has the odor of rotten eggs. Sometimes found in petroleum, it causes the foul smell of petroleum fractions. In dangerous concentrations, it is extremely corrosive and poisonous, causing damage to skin, eyes, breathing passages, and lungs and attacking and paralyzing the nervous system, particularly that part controlling the lungs and heart. In large amounts, it deadens the sense of smell. Also called hepatic gas or sulfureted hydrogen.

hydrogen sulfide cracking *n*: a type of corrosion that occurs when metals are exposed to hydrogen sulfide gas; it is characterized by minute cracks that form just under the metal's surface.

I

IADC *abbr*: International Association of Drilling Contractors.

impermeable *adj*: preventing the passage of fluid. A formation may be porous yet impermeable if there is an absence of connecting passages between the voids within it. See *permeability*.

inertia welding *n*: see *friction welding*.

inside diameter (ID) *n*: distance across the interior of a circle, especially in the measurement of pipe. See *diameter*.

internal-external upset end *n*: on tubing, casing, or drill pipe, the thickening at each end of the joint such that the internal diameter and the external diameter of the joint increase in thickness at each end. Compare *external upset end*, *internal upset end*.

internal upset end (IUE) *n*: on tubing, casing, or drill pipe, the thickening at each end of the joint such that the internal diameter of the joint increases in thickness at each end while the outside diameter remains unaffected. Compare *external upset end*, *internal-external upset end*.

International Association of Drilling Contractors (IADC) *n*: an organization of drilling contractors that sponsors or conducts research on education, accident prevention, drilling technology, and other matters of interest to drilling contractors and their employees. Its official publication is *The Drilling Contractor*. Address: Box 4287; Houston, TX 77210; (281) 578-7171; fax (281) 578-0589.

J

jar *n*: a percussion tool operated manually or hydraulically to deliver a heavy downward blow to fish stuck in the borehole. *v*: to apply a heavy blow to the drill stem by use of a jar or bumper sub.

joint *n*: in drilling, a single length (from 16 feet to 45 feet, or 5 metres to 14.5 metres, depending on its range length) of drill pipe, drill collar, casing, or tubing that has threaded connections at both ends. Several joints screwed together constitute a stand of pipe.

joint identifier *n*: a gauge for determining whether the connections of drill collars and tool joints match.

K

kelly *n*: the heavy steel tubular device, three- , four- , six- , or eight-sided, suspended from the swivel through the rotary table and connected to the top joint of drill pipe to turn the drill stem as the rotary table turns. It has a bored passageway that permits fluid to be circulated into the drill stem and up the annulus, or vice versa. Kellys manufactured to API specifications are available only in four- or six-sided versions, are either 40 or 54 feet (12 to 16 metres) long, and have diameters as small as 2½ (6 centimetres) and as large as 6 inches (15 centimetres).

kelly bushing (KB) *n*: a special device placed around the kelly that mates with the kelly flats and fits into the master bushing of the rotary table. The kelly bushing is designed so that the kelly is free to move up or down through it. The bottom of the bushing may be shaped to fit the opening in the master bushing or it may have pins that fit into the master bushing. In either case, when the kelly bushing is inserted into the master bushing and the master bushing is turned, the kelly bushing also turns. Since the kelly bushing fits onto the kelly, the kelly turns, and since the kelly is made up to the drill stem, the drill stem turns. Also called the drive bushing.

kilogram *n*: the metric unit of mass equal to 1,000 grams.

L

lay down pipe *v*: to pull drill pipe or tubing from the hole and place it in a horizontal position on a pipe rack. Compare *set back*.

lead tongs (pronounced "leed") *n pl*: the pipe tongs suspended in the derrick or mast and operated by a wireline connected to the breakout cathead. In coming out of the hole, they are used on the pin end of the joint for breaking out. In going into the hole, they are used on the box end as backup to the makeup tongs. In some areas, personnel call the makeup tongs the lead tongs if pipe is going into the hole; similarly, they call the breakout tongs the lead tongs if pipe is coming out of the hole.

lifting nipple *n*: a short piece of pipe with a pronounced upset, or shoulder, on the upper end, screwed into drill pipe, drill collars, or casing to provide a positive grip for the elevators. Also called a lifting sub or a hoisting plug.

lifting sub *n*: also called hoisting plug or lifting nipple. See *lifting nipple*.

lost pipe *n*: drill pipe, drill collars, tubing, or casing that has become separated in the hole from the part of the pipe reaching the surface, necessitating its removal before normal operations can proceed, i.e., a fish.

lubricant *n:* a substance—usually petroleum-based—that is used to reduce friction between two moving parts.

M

m *sym*: metre.

make a connection *v*: to attach a joint of drill pipe onto the drill stem suspended in the wellbore to permit deepening the wellbore by the length of the joint added (about 30 feet, or 9 metres).

make a trip *v*: to hoist the drill stem out of the wellbore to perform one of a number of operations, such as changing bits or taking a core, and then to return the drill stem to the wellbore.

make hole *v*: to deepen the hole made by the bit, i.e., to drill ahead.

make up *v*: 1. to assemble and join parts to form a complete unit (e.g., to make up a string of drill pipe). 2. to screw together two threaded pieces. 3. to mix or prepare (e.g., to make up a tank of mud). 4. to compensate for (e.g., to make up for lost time).

make up a joint *v*: to screw a length of pipe into another length of pipe.

makeup cathead *n*: a device that is attached to the shaft of the drawworks and used as a power source for screwing together joints of pipe. It is usually located on the driller's side of the drawworks. Also called spinning cathead. See *cathead*.

makeup shoulder *n:* the shoulder of a tool joint or drill collar that is connected to, or made up to, another tool joint or collar.

makeup tongs *n pl*: tongs used for screwing one length of pipe into another for making up a joint. See *lead tongs, tongs*.

male connection *n*: a pipe, coupling, or tool that has threads on the outside so that it can be joined to a female connection.

metre (m) *n*: the fundamental unit of length in the metric system. It is equal to about 3.28 feet, 39.37 inches, or 100 centimetres.

mg *abbr*: milligram.

millimetre *n*: a measurement unit in the metric system equal to 10^{-3} metre (0.001 metre). It is used to measure pipe and bit diameter, nozzle size, liner length and diameter, and cake thickness.

minimum tensile strength *n*: a specific value at which drill pipe will snap, or pull apart.

minimum yield strength *n*: a specific value at which drill pipe will permanently distort.

mm *abbr*: millimetre.

mousehole *n*: an opening through the rig floor, usually lined with pipe, into which a length of drill pipe is placed temporarily for later connection to the drill string.

mousehole connection *n*: the procedure of adding a length of drill pipe or tubing to the active string. The length to be added is placed in the mousehole, made up to the kelly, then pulled out of the mousehole and subsequently made up into the string.

mud *n*: the liquid circulated through the wellbore during rotary drilling and workover operations. In addition to its function of bringing cuttings to the surface, drilling mud cools and lubricates the bit and drill stem, protects against blowouts by holding back subsurface pressures, and deposits a mud cake on the wall of the borehole to prevent loss of fluids to the formation. Although it was originally a suspension of earth solids (especially clays) in water, the mud used in modern drilling operations is a more complex, three-phase mixture of liquids, reactive solids, and inert solids. The liquid phase may be fresh water, diesel oil, or crude oil and may contain one or more conditioners. See *drilling fluid*.

mud circulation *n*: the process of pumping mud downward to the bit and back up to the surface in a drilling or workover operation.

N

nipple *n*: a short, threaded tubular coupling used for making connections between pipe joints and other tools.

notch fatigue *n*: metal fatigue concentrated by surface imperfection, either mechanical (such as a notch) or metallurgical (such as a defect in the metal itself).

O

oil-base mud *n*: a drilling or workover fluid in which oil is the continuous phase and which contains from less than 2 percent and up to 5 percent water. This water is spread out, or dispersed, in the oil as small droplets. Compare *water-base mud*.

operating company *n*: see *operator*.

operator *n*: the person or company, either proprietor or lessee, actually operating an oilwell or lease, generally the oil company that engages the drilling contractor.

outside diameter (OD) *n*: the distance across the exterior circle, especially in the measurement of pipe. See *diameter*.

oxide *n*: a chemical compound in which oxygen is joined with a metal or a nonmetal.

P

packed bottomhole assembly *n*: a bottomhole assembly consisting of stabilizers and large-diameter drill collars arranged in a particular configuration to maintain drift angle and direction of a hole. This assembly is often necessary in crooked hole country. See *crooked hole country*.

pendulum assembly *n*: a bottomhole assembly composed of a bit and several large-diameter drill collars and stabilizers placed to allow the bottom drill collar to bend toward the vertical. The assembly works on the principle of the pendulum effect and is used to decrease drift angle. See *pendulum effect*.

pendulum effect *n*: the tendency of the drill stem—bit, drill collars, drill pipe, and kelly—to hang in a vertical position due to the force of gravity.

permeability *n*: 1. a measure of the ease with which a fluid flows through the connecting pore spaces of rock or cement. The unit of measurement is the millidarcy. 2. fluid conductivity of a porous medium. 3. ability of a fluid to flow within the interconnected pore network of a porous medium.

permeable *adj*: allowing the passage of fluid. See *permeability*.

petroleum *n:* a substance occurring naturally in the earth in solid, liquid, or gaseous state and composed mainly of mixtures of chemical compounds of carbon and hydrogen, with or without other nonmetallic elements such as sulfur, oxygen, and nitrogen. In some cases, especially in the measurement of oil and gas, petroleum refers only to oil—a liquid hydrocarbon—and does not include natural gas or gas liquids such as propane and butane. The API Measurement Coordination Department prefers that petroleum mean crude oil and not natural gas or gas liquids.

pH *abbr*: an indicator of the acidity or alkalinity of a substance or solution, represented on a scale of 0–14, 0–6.9 being acidic, 7 being neither acidic nor basic (i.e., neutral), and 7.1–14 being basic. These values are based on hydrogen ion content and activity.

pH value *n*: a unit of measure of the acid or alkaline condition of a substance. A neutral solution (such as pure water) has a pH of 7; acid solutions are less than 7; basic, or alkaline, solutions are more than 7. The pH scale is a logarithmic scale. A substance with a pH of 4 is more than twice as acid as a substance with a pH of 5. Similarly, a substance with a pH of 9 is more than twice as alkaline as a substance with a pH of 8.

pin *n*: the male section of a tool joint or drill collar.

pipe *n*: a long, hollow cylinder, usually steel, through which fluids are conducted. Oilfield tubular goods are casing (including liners), drill pipe, tubing, or line pipe.

pipe protector *n*: prevents drill pipe from rubbing against the hole or against the casing.

pipe rack *n*: a horizontal support for tubular goods.

pipe tongs *n pl*: see *tongs*.

porosity *n*: 1. the condition of being porous (such as a rock formation). 2. the ratio of the volume of empty space to the volume of solid rock in a formation, indicating how much fluid a rock can hold.

power slips *n pl*: see *slips*.

power tongs *n pl*: a wrench that is used to make up or break out drill pipe, tubing, or casing on which the torque is provided by air or fluid pressure. Conventional tongs are operated by mechanical pull provided by a jerk line connected to a cathead.

preload *n*: the amount of torque required to makeup two tool joints or two drill collars.

pressure *n*: the force that a fluid (liquid or gas) exerts uniformly in all directions within a vessel, pipe, hole in the ground, and so forth, such as that exerted against the inner wall of a tank or that exerted on the bottom of the wellbore by a fluid. Pressure is expressed in terms of force exerted per unit of area, as pounds per square inch, or in kilopascals.

pressure coupling *n*: 1. a joining device, such as a Squnch joint, in which the force of weight is used to make the connection. 2. a pipe or line coupling that is capable of maintaining a tight, leak-free connection under high pressures.

psi *abbr*: pounds per square inch.

psia *abbr*: pounds per square inch absolute.

psi/ft *abbr*: pounds per square inch per foot.

psi/g *abbr*: pounds per square inch gauge.

pup joint *n*: a length of drill or line pipe, tubing, or casing considerably shorter than 30 feet (9 metres).

pure fatigue *n*: metal fatigue for which no cause can be determined.

R

rack *n*: framework for supporting or containing a number of loose objects, such as pipe. See *pipe rack*.

rack pipe *v*: 1. to place pipe withdrawn from the hole on a pipe rack. 2. to stand pipe on the derrick floor when pulling it out of the hole.

range length *n*: a grouping of pipe lengths. API designation of range lengths is as follows:

	Range 1 (ft/m)	**Range 2 (ft/m)**	**Range 3 (ft/m)**
Casing	16–25 feet 5–1.5 metres	25–34 feet 1.5–10 metres	34–48 feet 10–14.6 metres
Drill pipe	18–22 feet 5.5–6.7 metres	27–30 feet 8.2–9.1 metres	38–45 feet 11.6–13.7 metres
Tubing	20–24 feet 6.1–7.3 metres	28–32 feet 8.5–9.8 metres	

rathole *n*: 1. a hole in the rig floor, 30 to 35 feet (9 to 11 metres) deep, which is lined with casing that projects above the floor and into which the kelly and swivel are placed when hoisting operations are in progress. 2. a hole of a diameter smaller than the main hole and drilled in the bottom of the main hole. *v*: to reduce the size of the wellbore and drill ahead.

rathole connection *n*: the addition of a length of drill pipe or tubing to the active string using the rathole instead of the mousehole, which is the more common connection. The length to be added is placed in the rathole, made up to the kelly, pulled out of the rathole, and made up into the string.

reactive torque *n*: the tendency of the drill string to turn in a direction opposite that of the bit, a factor for which a driller must compensate when using a downhole motor.

reface *v*: to renew or smooth a faced surface by recutting or regrinding.

rig *n*: the derrick or mast, drawworks, and attendant surface equipment of a drilling or workover unit.

rig crew member *n*: see *rotary helper*.

rig floor *n*: the area immediately around the rotary table and extending to each corner of the derrick or mast, that is, the area immediately above the substructure on which the drawworks, rotary table, and so forth rest. Also called derrick floor, drill floor.

rig manager *n*: an employee of a drilling contractor who is in charge of the entire drilling crew and the drilling rig. Also called toolpusher, drilling foreman, rig supervisor, or rig superintendent.

rig superintendent *n*: see *toolpusher*.

rig supervisor *n*: see *toolpusher*.

rotary *n*: the machine used to impart rotational power to the drill stem while permitting vertical movement of the pipe for rotary drilling. Modern rotary machines have a special component, the rotary or master bushing, to turn the kelly bushing, which permits vertical movement of the kelly while the stem is turning.

rotary drilling *n*: a drilling method in which a hole is drilled by a rotating bit to which a downward force is applied. The bit is fastened to and rotated by the drill stem, which also provides a passageway through which the drilling fluid is circulated. Additional joints of drill pipe are added as drilling progresses.

rotary helper *n*: a worker on a drilling or workover rig, subordinate to the driller, whose primary work station is on the rig floor. On rotary drilling rigs, there are at least two and usually three or more rotary helpers on each crew. Sometimes called floorhand, floorman, rig crew member, or roughneck.

rotary-shouldered connection *n*: the threaded and shouldered joint used in rotary drilling to join the various components of the drill stem.

rotary speed *n*: the speed, measured in rpm, at which the rotary table is operated.

rotary table *n*: the principal component of a rotary, or rotary machine, used to turn the drill stem and support the drilling assembly. It has a beveled gear arrangement to create the rotational motion and opening into which bushings are fitted to drive and support the drilling assembly.

rotary torque *n*: the rotational force applied to turn the drill stem.

S

saver sub *n*: a device made up in the drill stem to absorb much of the wear between frequently broken joints (such as between the kelly and the drill pipe).

seamless drill pipe *n*: drill pipe that is manufactured in one continuous piece. Most drill pipe is of seamless construction.

set back *v*: to place stands of drill pipe and drill collars in a vertical position to one side of the rotary table in the derrick or mast of a drilling or workover rig. Compare *lay down pipe.*

shoulder *n*: 1. the flat portion machined on the base of the bit shank that meets the shoulder of the drill collar and serves to form a pressure-tight seal between the bit and the drill collar. 2. the flat portion of the box end and the pin end of a tool joint; the two shoulders meet when the tool joint is connected and form a pressure-tight seal.

shoulder dressing tool *n*: in drilling, an instrument used to cut or grind the shoulder of a collar or tool joint until it is smooth or flat.

single *n*: a joint of drill pipe. Compare *double, thribble,* and *fourble.*

slant-hole rig *n*: a drilling rig used to drill directional wells. See *directional drilling.*

slips *n pl*: wedge-shaped pieces of metal with teeth or other gripping elements that are used to prevent pipe from slipping down into the hole or to hold pipe in place. Rotary slips fit around the drill pipe and wedge against the master bushing to support the pipe. Power slips are pneumatically or hydraulically actuated devices that allow the crew to dispense with the manual handling of slips when making a connection. Packers and other downhole equipment are secured in position by slips that engage the inner surface of casing.

sour corrosion *n*: embrittlement and subsequent wearing away of metal caused by contact of the metal with hydrogen sulfide.

sour crude *n*: see *sour crude oil.*

sour crude oil *n*: oil containing hydrogen sulfide or another acid gas.

sour gas *n*: gas containing an appreciable quantity of hydrogen sulfide and/or mercaptans.

sour hole *n*: a wellbore or formation known to contain hydrogen sulfide gas.

spinning chain *n*: a Y-shaped chain used to spin up (tighten) one joint of drill pipe into another. One end of the chain is attached to the tongs, another end to the spinning cathead, and the third end left free. The free end is wrapped around the tool joint, and the cathead pulls the chain off the joint, causing the joint to spin rapidly and tighten up. After the free end of the chain is pulled off the joint, the tongs are secured in the spot vacated by the chain and continued pull on the chain (and thus on the tongs) by the cathead makes up the joint to final tightness.

spinning wrench *n*: air-powered or hydraulically powered wrench used to spin drill pipe in making or breaking connections.

spirally grooved drill collar *n*: a drill collar with a round cross section that has a long continuous groove or flute machined helically into its outer surface. The spiraled groove provides space between the wall of the hole and the body of the collar, minimizing the area of contact between the hole wall and the collar; thus the possibility of differential pressure sticking is reduced.

square drill collar *n*: a special drill collar, square but with rounded edges, used to control the straightness or direction of the hole; often part of a packed-hole assembly.

stab *v*: to guide the end of a pipe into a coupling or tool joint when making up a connection.

stabilizer *n*: a tool placed on a drill collar near the bit that is used, depending on where it is placed, either to maintain a particular hole angle or to change the angle by controlling the location of the contact point between the hole and the collars.

stand *n*: the connected joints of pipe racked in the derrick or mast during a trip. The usual stand is about 90 feet (about 27 metres) long, which is three lengths of drill pipe screwed together (a thribble).

stiff drilling assembly *n*: see *packed bottomhole assembly.*

stress *n*: a force that, when applied to an object, distorts or deforms it.

stress concentrator *n*: a notch or pit on a pipe or joint that raises the stress level and concentrates the breakdown of the metal structure. Also called a stress riser.

stress riser *n*: see *stress concentrator.*

string *n*: the entire length of casing, tubing, sucker rods, or drill pipe run into a hole.

stringer *n*: an extra support placed under the middle of racked pipe to keep the pipe from sagging.

stuck pipe *n*: drill pipe, drill collars, casing, or tubing that has inadvertently become immovable in the hole. Sticking may occur when drilling is in progress, when casing is being run in the hole, or when the drill pipe is being hoisted.

stuck point *n*: the depth in the hole at which the drill stem, tubing, or casing is stuck.

sub *n*: a short, threaded piece of pipe used to adapt parts of the drilling string that cannot otherwise be screwed together because of differences in thread size or design. A sub (i.e., a substitute) may also perform a special function. Lifting subs are used with drill collars to provide a shoulder to fit the drill pipe elevators; a kelly saver sub is placed between the drill pipe and the kelly to prevent excessive thread wear of the kelly and drill pipe threads; a bent sub is used when drilling a directional hole.

sulfide stress cracking *n*: a form of hydrogen embrittlement. It involves stress and the absorption of hydrogen in the presence of hydrogen sulfide. Sulfide stress cracking is a frequent cause of drill stem failure.

sweet corrosion *n*: the deterioration of metal caused by contact with carbon dioxide in water.

swelled box *n*: a box connection on a tool joint that has been belled by too much torque.

T

take out *v*: to remove a joint or stand of pipe from the drill stem.

tapered string *n*: drill pipe, tubing, sucker rods, and so forth with the diameter near the top of the well larger than the diameter below.

tensile *adj*: of or relating to tension.

tensile load *n*: the amount of longitudinal stress borne by a substance.

tensile strength *n*: the greatest longitudinal stress that a metal can bear without tearing apart. A metal's tensile strength is greater than its yield strength.

tensile stress *n*: stress developed by a material bearing a tensile load. See *stress*.

tension *n*: the condition of a string, wire, pipe, or rod that is stretched between two points.

thread *n*: a continuous helical rib, as on a screw or pipe.

thread compound *n*: see *dope*.

thread dope *n*: see *dope*.

threaded coupling *n*: a type of connector that has threads on each end, making it possible to screw two pieces of pipe together. Compare *pressure coupling*. See also *coupling*.

thread profile gauge *n*: a device to measure the amount of wear or stretch on pipe threads.

thread protector *n*: a metal or plastic device that is screwed onto or into pipe threads to protect them from damage when the pipe is not in use.

thribble *n*: a stand of pipe made up of three joints and handled as a unit. Compare *double*, *fourble*, *single*.

thribble board *n*: the name used for the derrickhand's working platform, the monkeyboard, when it is located at a height in the derrick equal to three lengths of pipe joined together.

throw the chain *v*: to flip the spinning chain up from a tool joint box so that the chain wraps around the tool joint pin after it is stabbed into the box. The stand or joint of drill pipe is turned or spun by a pull on the spinning chain from the cathead on the drawworks.

tong area *n*: the area where tongs are attached to the drill pipe and drill collars during makeup and breakout. On the drill string, the tong area is the tool joint rather than the drill pipe.

tong dies *n pl*: very hard and brittle pieces of serrated steel that are installed in the tongs and that grip or bite into the tool joint of drill pipe when the tongs are latched onto the pipe.

tongman *n*: the member of the drilling crew who handles the tongs.

tongs *n pl*: the large wrenches used to make up or break out drill pipe, casing, tubing, or other pipe; variously called casing tongs, pipe tongs, and so forth, according to the specific use. Power tongs are pneumatically or hydraulically operated tools that serve to spin the pipe up tight and, in some instances, to apply the final makeup torque.

tool joint *n*: a heavy coupling element for drill pipe. It is made of special alloy steel and has coarse, tapered threads and seating shoulders designed to sustain the weight of the drill stem, withstand the strain of frequent coupling and uncoupling, and provide a leakproof seal. The male section of the joint, or the pin, is attached to one end of a length of drill pipe, and the female section, or box, is attached to the other end. The tool joint may be welded to the end of the pipe, screwed on, or both. A hard-metal facing is often applied in a band around the outside of the tool joint to enable it to resist abrasion from the walls of the borehole.

toolpusher *n*: an employee of a drilling contractor who is in charge of the entire drilling crew and the drilling rig. Also called a drilling foreman, rig manager, rig superintendent, or rig supervisor.

top drive *n*: a device similar to a power swivel that is used in place of the rotary table to turn the drill stem. It also suspends the drill stem in the hole and includes power tongs. Modern top drives combine the elevator, tongs, swivel, and hook.

torque *n*: the turning force that is applied to a shaft or other rotary mechanism to cause it to rotate or tend to do so. Torque is measured in units of length and force (foot-pounds, newton-metres).

torsion *n*: twisting deformation of a solid body about an axis in which lines that were initially parallel to the axis become helices. Torsion is produced when part of the pipe turns or twists in one direction while the other part remains stationary or twists in the other direction.

torsional yield strength *n*: the amount of twisting force that a pipe can withstand before twisting off.

toughness *n*: measure of steel's ability to withstand a crack or flaw without fracturing. It can be altered by changing chemical composition, microstructure, and heat treatment. Also called ductility. Compare *brittleness*.

transition zone *n*: the first few joints of drill string above the drill collars.

trip *n:* the operation of hoisting the drill stem from and returning it to the wellbore. *v*: shortened form of "make a trip."

trip in *v*: see *go in the hole*.

trip out *v*: see *come out of the hole*.

tripping *n*: the operation of hoisting the drill stem out of and returning it to the wellbore. See *make a trip*.

tubing *n:* relatively small-diameter pipe that is run into a well to serve as a conduit for the passage of oil and gas to the surface.

tubular goods *n pl*: any kind of pipe. Oilfield tubular goods include tubing, casing, drill pipe, and line pipe. Also called tubulars.

turn to the right *v*: on a rotary rig, to rotate the drill stem clockwise. When drilling ahead, the expression "on bottom and turning to the right" indicates that drilling is proceeding normally.

twistoff *n*: a complete break in pipe caused by metal fatigue. *v*: to break something in two or to break apart, such as the head of a bolt or the drill stem.

U

ultimate strength *n*: the greatest stress that a substance can stand under normal short-term experiments.

upper string *n*: any part of the drill stem, tubing string, or casing string that lies in the upper part of the borehole.

upset *n*: thickness forged to the end of a tubular (such as drill pipe) to give the end extra strength. *v*: to forge the ends of tubular products so that the pipe wall acquires extra thickness and strength near the end. Upsetting is usually performed to provide the thickness needed to form threads so that the tubular goods can be connected.

upset runout *n*: the point on a joint of drill string where the upset ends and the nominal wall of the pipe meet.

V

V-door *n*: an opening at floor level in a side of a derrick or mast. The V-door is opposite the drawworks and is used as an entry to bring in drill pipe, casing, and other tools from the pipe rack. The name comes from the fact that on the old standard derrick, the shape of the opening was an inverted V.

W

wall cake *n*: also called filter cake or mud cake. See *filter cake*.

wandering *n*: the tendency of the drill bit to deviate horizontally parallel to tilted strata.

washout *n*: 1. excessive wellbore enlargement caused by solvent and erosional action of the drilling fluid. 2. a fluid-cut opening caused by fluid leakage.

water-base mud *n*: a drilling mud in which the continuous phase is water. In water-base muds, any additives are dispersed in the water. Compare *oil-base mud*.

weight on bit (WOB) *n*: the amount of downward force placed on the bit by the weight of the drill collars.

well *n*: the hole made by the drilling bit, which can be open, cased, or both. Also called borehole, hole, or wellbore.

wellbore *n*: a borehole; the hole drilled by the bit. A wellbore may have casing in it or it may be open (uncased); or part of it may be cased, and part of it may be open. Also called a borehole or hole.

wiper *n*: a circular rubber device with a split in its side that is put around drill pipe to wipe or clean drilling mud off the outside of the pipe as the pipe is pulled from the hole.

WOB *abbr*: (pronounce each letter): weight on bit.

wobble *n*: movement between the mating surfaces of box and pin in a tool joint. *v*: to move in a rocking motion.

Y

yield point *n*: the maximum stress that a solid can withstand without undergoing permanent deformation either by plastic flow or by rupture. See *tensile strength*.

yield strength *n*: a measure of the force needed to deform tubular goods to the extent that they are permanently distorted.

yield value *n*: the resistance to initial flow, or the stress required to start fluid movement. This resistance is caused by electrical charges located on or near the surfaces of the particles. The values of the yield point and thixotropy, respectively, are measurements of the same fluid properties under dynamic and static states. The Bingham yield value, reported in pounds/100 square feet, is determined from a direct-indicating viscometer by subtracting the plastic viscosity from the 300-rpm reading. Also called yield point.

Z

zone *n*: a rock stratum that is different from or distinguished from another stratum (e.g., a pay zone).